ROBOT ZOMBIES

Transhumanism and the Robot Revolution

Xaviant Haze
and
Estrella Eguino

Adventures Unlimited Press

Robot Zombies

by Xaviant Haze and Estrella Eguino

ISBN 978-1-939149-51-0

Published by:
Adventures Unlimited Press
One Adventure Place
Kempton, Illinois 60946 USA
auphq@frontiernet.net

www.AdventuresUnlimitedPress.com

ROBOT ZOMBIES

Transhumanism and the Robot Revolution

Xaviant Haze
and
Estrella Eguino

Adventures Unlimited Press

TABLE OF CONTENTS

Also by Xaviant Haze

•Elvis Lives!
•Aliens in Ancient Egypt
•The Suppressed History of Ancient America

http://xavianthaze.blogspot.com

Estrella Eguino can be reached at:

http://about.me/estrellaeguino.com
http://estrellaeguino.blogspot.com

ROBOT
ZOMBIES

Transhumanism and
the Robot Revolution

Xaviant Haze
and
Estrella Eguino

FOREWORD
by
Dr. Roman V. Yampolskiy

There are certain topics everyone loves. Robots is one such theme, zombies is another. Put them together, sprinkle some conspiracy theory, Hollywood blockbusters and sex with spiritual robots and you get a killer book. *Robot Zombies* written by Xaviant Haze and Estrella Eguino is written from the point of view of a curious fan of technology trying to understand where technology is going by referring to historical background, association with what we see in the mass media (movies, TV shows, news) and making certain general assumptions. It is not a boring scholarly work, but a very fascinating reading for someone interested in transhumanism and robot revolution.

I recommend reading this book to anyone interested in what the future is going to be like. Personally, I would be very surprised if by 2045 we don't have superintelligent machines which will to us appear to operate and produce novel technology at the speed close to instantaneous, hence Technological Singularity. This is a direct consequence of Moore's law continuing as predicted, perhaps encompassing multiple paradigms such as quantum computing. In order to be that intelligent and self-improving a system has to be able to make novel, unpredictable decisions. It has to be independent. If we could precisely predict decisions it would make we would be as intelligent as the said system.

Machines have long ago begun to replace humans working in physical labor-related jobs. Next, we are going to see a great increase in the number of "intellectual jobs" which could be done by machines. Eventually, as AIs reach a level of human performance all jobs will be automated. This will produce a very significant change is how society is organized, our economy, class structure

i

and such concepts as money. I am not too worried about auto-mation of labor; it is the next step in this process, which is really dangerous. Very shortly after getting to human-level performance machines will self-improve to levels far beyond—superintelligence. Consequently, I am more concerned about machines taking our place in the universe, and not as concerned about them taking our place in the factories. Now, how to deal with that problem is a much more complicated issue, I don't think anyone has a solution to it right now. The work on it is just beginning.

The main problem with technology exceeding human capabilities is of course the control problem. How can we control an entity smarter than us? An entity which is capable of deriving its own plans, which may not necessarily be well aligned with our values. I see a non-zero probability of such machines being extremely dangerous to all life because of either poor goal alignment or poor implementation. Enjoy Robot Zombies, and perhaps it will inspire you to start thinking about some of the most interesting questions facing humanity.

September 26, 2015

Dr. Roman V. Yampolskiy
Author of *Artificial Superintelligence: A Futuristic Approach*

INTRODUCTION

The technological leaps made in the fields of programming, artificial intelligence, bioengineering and nanotechnology have thrust humans into a fantastical present day that far surpasses the visions of all the great sci-fi authors who dazzled our imaginations with fictional views of scientifically possible futures. That future is now—a world of zombies, robots and artificial intelligence is breaking into the mainstream. Most of the population is too busy playing Candy Crush to heed the publicized warnings from physicist Stephen Hawking and entrepreneur Elon Musk regarding the development of AI that will learn how to replicate itself and be more dangerous to humans than our own vast arsenal of nukes. We can assume these men, both futurist visionaries, have already calculated the inevitable moment in time when machines reach a point of complete self-awareness—a state that humans, in their great numbers, have failed to accomplish and surpass humans in intelligence. The moment Hawking and Musk are fearful of is called 'the singularity.' Are they correct in predicting our end? According to Hawking, self-aware artificial intelligence would "take off on its own and re-design itself at an ever-increasing rate, and humans, who are limited by slow biological evolution, wouldn't compete and would be superseded." How close are we to singularity? In late 2012, the Japanese robot QBO, using stereoscopic vision, was able to recognize itself in a mirror.

Science and technology are ever pushing forward creating new modes of living and transforming life as we know it, perhaps even giving humans a shot at immortality. This Transhumanist future has been the goal of the global elite, who have knowingly steered mankind with social and bioengineering programs for the past 200 years in the process to bring it about. In man's quest to become immortal, a new line has been drawn in the sand; one that

1

threatens to replace the human spirit with artificial reactions. Our daily routines already involve automation of some form. From cars and cookies, to computers, our lives are replete with the work of robotic applications. The lightning speed advances in technology arise so quickly and in such great numbers that we can actually see the exponential growth in artificial intelligence happening in front of us. The human brain doesn't operate at full capacity and we don't know why; this is why we're becoming obsolete. The traditional role of an accomplished, steady-handed surgeon in an operating room is destined to become one of a skilled computer operator performing the same procedure from his office via robotic camera and the actual surgical apparatus set up where the patient is. Where do robots come from? Where are they and where can we find them?

It was the visionary author Isaac Asimov who introduced robots to pop culture. Born in Russia and raised in America, Asimov coined the term 'robotic' and authored over 500 books and an estimated 90,000 letters, many of which brought fictional robots in touch with human interaction. Asimov is considered one of the most prolific scribes of all time, and he began writing about various forms of humanoid robots at the age of 19 leading up to his classic *I, Robot*, which brought the idea of robots living alongside humans into the mainstream. Before Asimov saw the 'robot' with metals and wires the term had been used by Czech playwright Josef Capek, who associated 'robota' with the lifeless, mindless servitude of the peasant. This 'robot' ended up in Karel Capek's (Josef's brother) play *Rossum's Universal Robots* that premiered in New York City in 1922 and was translated and published in English a year later. But the idea of the robot or mechanical automaton has existed since ancient times and we find, for example, Egyptian and Greek myths filled with stories of mechanical automation.

Our modern robotic world was shaped by the business foresight of Joseph Engelberger, the man who sold the first robot for industrial purposes to the General Motors Corporation in 1961. Fast forward 51 years to 2012 when Amazon purchased the robotic manufacturing company Kiva Systems for 775 million dollars, with the aim to eventually replace all human warehouse workers with robots. The world is not going back to sticks and stones, and at this point unless there's a catastrophe of event horizon proportions we can pretty much be assured robotics is here to transform humans and human behavior. Other scientists, speared primarily by Stephen Hawking, have expressed concern about the dangers that could arise with overuse of decision-making robots that may be able to create goals and objectives and work towards achieving them. This is of special concern, particularly in the military, where automation and advanced weaponry is used in clandestine operations. Can militarized machines truly rise above their station and violate the first law of robotics? What is real, and what is hype?

This book will examine the history and the future of robotics, artificial intelligence and a Trans-humanist utopia/ dystopia, merging man with machine. How did it all begin, and what can we expect in the present, the near future and the distant future? Examining the fascinating role of robotics and artificial intelligence from a practical human perspective in our everyday world we discover the mind-altering process necessary to accept and integrate with the inevitable. Scientific developments and advancements in the field of technology appear to generally be exposed first in film. Motion pictures are personal experiences created to 'put you there.' Consequently films have the deepest reach into individual consciousness. Make no mistake, those Hollywood blockbuster sci-fi and fantasy movies have several purposes. This type of expression is referred to in metaphysical circles as 'the craft' and it is designed to speak to those within the masses who are aware. What has Hollywood been serving?

Robot Zombies

1.
To Live and AI in Hollywood

Before seeing life on the screen, science fiction was alive and well living in the radio, and in literature from all parts of the world, including the Bible. The onset of silent film in the 1920's brought classic tales of fantasy and futurism to the screen, even while radio was the mass medium. Among the flood of sci-fi titles that emerged during this time, with themes of technology and robotics, was the classic 1927 German film *Metropolis*. With futuristic sets designed by famed Italian architect Antonio Sant'Elia and directed by Fritz Lang from a story written by his wife, *Metropolis* is considered the pioneering epic that launched the science fiction genre. The movie touches the theme of robotics, along with class separation, as the lead female role is abducted. A robot likeness replaces her and begins causing trouble for the male workers of the city, whom she instigates into revolt. In the end the robot is burned at the stake and its identity revealed when the skin peels away. The message in this movie clearly associates the robot clone in a leadership capacity over the working masses that end up 'killing' technology. The special effects used in the movie were startling for the time, and to this day represent a stroke of quality and ingenuity. It is worthwhile mentioning that Fritz Lang had a personal interest in science fiction that included the friendship of Hermann Oberth, who worked as an advisor on Lang's *Woman in the Moon* and attempted to build an actual working rocket to be used in the film. Wernher von Braun, considered the leading pioneer of rocket science, was a German aerospace engineer and space architect member of the Nazi party and the SS. He was a member of the Verein fur Raumschiffahrt (Society for Space Travel), a German rocket

5

science association that included members from other countries and was founded by world-renowned rocket scientists Johannes Winkler and Max Valier. It is around this time that we begin to see the interest of advanced technology at the highest of levels targeted to mass appeal through film, an integration that exists today. It is interesting to note *Metropolis* received the highest financing for a film of its day at more than five million Reichsmarks, making it the most expensive silent film ever produced. However, the film was a commercial flop and its historical significance wouldn't be realized until decades later when several different versions surfaced. The closest to the original version is a copy found in Argentina with 25 additional minutes. It is not a German museum of art that holds these copies, but The Smithsonian.

Orson Welles' rendition of H.G. Wells' *War of the Worlds* is another classic example of the shifting mindset of the public, mostly younger generations, who were awakened by entertainment

Maria the Robot—*Metropolis*

to take notice of technology and glimpses of a scientifically frightening future. The infamous 1938 radiobroadcast simulating an alien invasion frightened people—mostly those who hadn't read the book. The result put Welles in front of the FCC, which in turn used the opportunity to gauge and duly note the power of mass communication. Media and government closely monitored this era of radio and film. The first sci-fi and fantasy Hollywood films were based on books and existing themes, such as the *Flash Gordon* series, but unlike its German competition, these films were not intended to deliver any particular message yet. It wasn't until what it is now referred to as 'The Golden Age of Sci-Fi,' ushered by the explosion of the atomic bomb and the Roswell incident, that the themes of space and time travel, alien beings and interplanetary invasions, robots and machines, suddenly became popular. Hollywood films that explored these themes were *Destination Moon, The Time Machine, War of the Worlds, Invasion of the Body Snatchers, The Angry Red Planet, The Thing From Another World*, and George Pal's *Conquest of Space*. These titles are among the most memorable. Then the movie *The Day the Earth Stood Still* came around and the world faced its nuclear fears head on. More sci-fi movies were made during the 50's than any other time, many of them campy, with unknown stars, corny dialogue and absurd special effects. The underlying messages in these films blossomed and it was evident that bigger budgets would mean bigger audiences. The messages remained more or less constant—man was heading to an unknown future where invasions from outer space and chemical mutations would destroy a huge portion of humanity before someone could figure out a simple way to stop them. In a nutshell: abstain from technological progress because it upsets the order of the classes. Then two movies were produced that again changed the scene and served the dual purpose of top-of-mind programming and technological playground.

The first was *Forbidden Planet* in 1956. The movie boasted technology and special effects not seen before in film. The plot, based on Shakespeare's *The Tempest,* involved space travel to an

Forbidden Planet—MGM

alien planet and the debut of "Robby the Robot," a mechanical assistant programmed with superior knowledge and apparently a distinct dry-wit personality. The movie is the first to introduce an all-electronic musical score.

Three years later *On the Beach*, based on the 1957 post-apocalyptic novel written by Nevil Shute featured a cast that would make anybody curious enough to go see it on the big screen. After *On the Beach* nobody has been able to look at a nuclear reactor without some form of trepidation. These movies are all a product of their time. High-end technology was in its infancy after WWII and filmmakers, particularly in Berlin and Hollywood, were only too eager to incorporate the remnants of advancing knowledge that made it out of the Nazi camp. The prevailing mentality was that of fear and defense. The climax of futuristic technology in film came at the end of the 1950's when Gene Roddenberry created *Star Trek* and imaginations everywhere leaped light years ahead of their time. Unlike the frenzy of low budget sci-fi "B" movies that swamped the years immediately following WWII and well into the end of the decade, the 1960's saw the genre go through a transforming change. The quality of sci-fi films became decidedly better and so did the efforts to maintain plausible open-ended outcomes, again reflecting the prevailing scientific mindset of the time.

The real race for space was going on and technological breakthroughs again saw their debut for the general public in film. *Planet of the Apes,* based on a novel by Pierre Boulle, pitted humans against a future reality where apes ruled over man as a result of misuse of technology. However, the most significant work from this period is *2001: A Space Odyssey,* directed by Stanley Kubrick. Kubrick wrote it in collaboration with Arthur C. Clarke. 'The Sentinel,' a short story that Clarke wrote in 1940 which topic transcended the ideas of the time, inspired it. The entanglement between Kubrick and the reportedly fake lunar landing of Apollo 11 is a complex subject and one that has been thoroughly exposed by qualified experts for anyone to review and analyze. It is not what we are concerned with here however. It is to be noted that *2001: A Space Odyssey* is a landmark film with an evolving message hinting at the conflict of computers controlling humans.

The film featured a sentient computer, HAL 9000 that in the end becomes man's adversary. In the satellite sequence of the film each satellite has an insignia. One has a German flag and a Maltese Cross; one shows the Chinese Air Force insignia as the shot scrolls up to the Moon and the other toward the rising Sun. The music score for this sequence is *The Blue Danube*, the name of England's first nuclear weapon. This scene is saying technology is a threat to humans. The movie makes specific references, both visually and in dialogue, that point to HAL being identified with IBM. For starters, each letter in HAL is the preceding letter for each letter in IBM. To further corroborate this, the letters IBM can be seen in several shots, including the computer display of the docking airship and on the buttons of Bowman's spacesuits. The clearest evidence of this association Kubrick was deliberately trying to make was when the letters IBM are seen reflecting the right way around across Dave's face. Right at that moment Dave asks, "Do you read me HAL?"

"Daisy Bell," the first song performed using a computerized speech synthesizer, is the song HAL sings as he is being shut down. This is a reference to the first demonstration of synthesized computer speech produced in 1962 by an IBM 704 computer at the Langley Research Facility in Urbana, Illinois where the computer sang Daisy Bell. In this scene HAL says Mr. Langley instructed him in a plant in Urbana, Illinois. The movie's special visual effects and stunning photography set a new benchmark for quality in film with a mesmerizing and almost hypnotic allure. Its accurate portrayal of space travel would raise the bar for future films if they were going to be taken seriously. Most importantly this film planted seeds and spoke to the public at a loftier level that

Stanley Kubrick directing Hal sequence from *2001* — Gary Green

transcends the scope of the story.

Sci-fi robots of films in the 1970s include the most famous robots in the world: C-3PO and R2-D2. In the most epic Sci-fi movie of all time, *Star Wars,* humans, aliens and machines coexist and fight against evil. The feel-good message of this Hollywood blockbuster pointed to a populated galaxy of the scale of *2001: A*

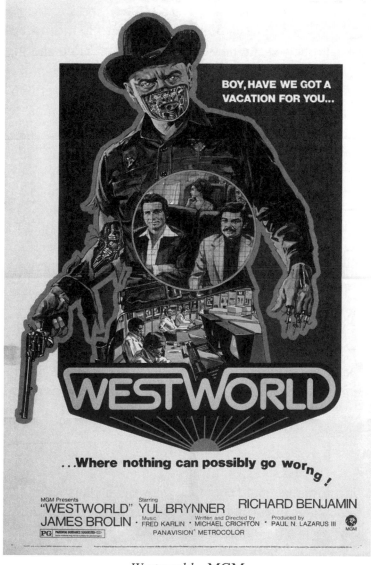

Westworld — MGM

Space Odyssey, but more Disney-style. In *Star Wars*, the message is machines are not to be feared, and robots are subservient to man and other beings—like in the *Stepford Wives*.

With relief from performing menial tasks, humans of the future can purchase fantasy vacation packages in the amusement park, Delos, in the sci-fi movie *Westworld,* which pitted man against robot. The film featured a relentless pursuit by a robot gunslinger after a computer malfunction causes all the robots in the park to kill the vacationing guests. Outside the spectrum of artificial and mechanical robots vs. humans in 1970s films, we find another type of adversary, brainwashing, no less influenced by technology, in the film *A Clockwork Orange*, also a Kubrick production. What sense does it make to create artificial, mechanical computers that behave like humans when humans can be brainwashed to behave like mindless robots?

In the 1980's sci-fi movies explored more robots, artificial intelligence, and alien territory. During a decade marked with the birth of the Internet and exponential space technology, the messages coming out of Hollywood with movies such as *Alien, Blade Runner, Scanners* and *Tron,* were as complex as the technology itself. There was an explosion of high-tech scenarios where humans were the hunted species at the mercy of deceit. In these films the message seemed to be that we were not alone and the beings we were sharing space with could enslave us, and possibly did. Then in 1984 when James Cameron introduced moviegoers to a futuristic view of the world dominated by computerized machines bent on destroying the human race, minds young and old alike went racing over the survival of mankind. Despite its stark premise, the movie *Terminator* became an instant blockbuster and cult classic. The film spawned additional sequels with groundbreaking special effects that exposed the world to metallic alloys that could chemically alter and reshape into various forms, and mechanical eyes that scanned humans and zoomed into molecular levels to read heat signatures.

Balancing the pessimistic outlook introduced by *Terminator*, subsequent genre movies of the 1990's, such as

Bicentennial Man, written by Isaac Asimov, and Steven Spielberg's *Artificial Intelligence* proposed an alternate outcome promoting the role of robots in the future, not as one to annihilate humanity, but to enhance it. Other films like *Lawnmower Man, Virtuosity, Johnny Mnemonic, Total Recall, Gattacca*, and *The Matrix* dealt more with threats to the network by humans. Movies like *The Iron Giant, Space Camp, Short Circuit, and Titan A.E.,* brought the theme to even more prominence and younger audiences.

Johnny 5—*Short Circuit 2*

The technology in these movies is astounding, including the rate at which programming has grown and the fine details of special effects. The message is clear: humanity can push back. These films were not Hollywood productions, but independent films that captured the public's attention, even over blockbuster productions of a recycled alien invasion in *Independence Day*, and an end of the world comet strike in *Armageddon*. The unique film entries of the new century explore the merging of mind and machine through movies such as *Vanilla Sky* and *The Eternal Sunshine of the Spotless Mind*, both relying on a mind-altering technology that allows a person to interact in a manufactured reality that does not exist. The *Caprica* miniseries is a good case in point

13

for the smaller screen of television. In *Minority Report,* the Tom Cruise film based on a futuristic short story by sci-fi legend Philip K. Dick, audiences were exposed for the first time to holographic information appearing on a one-dimensional screen. As more original entries in the field of robotics and technology bloomed, both from the powerhouses that lull the world to sleep, and those that struggle to awaken it, one thing is certain—if Hollywood puts millions behind a film, they want everyone's attention.

The film *Interstellar* is the most recent sci-fi entry that touches themes that are scientific fact. Its message is both disturbing and hopeful, placing emphasis on humans' superiority over machines. The fascination humans have with robots and technology goes back thousands of years. The mechanical and computing disciplines of technology were born to live; it appears, intrinsically integrated forever.

2.
Oh Robot, Where Art Thou?

It all started with Pygmalion, an ancient Greek legend based on an even older Phoenician myth. It told the story of a man-made sculpture that came to life after its maker fell in love with his creation and pleaded to a goddess to give his sculpture life. The idea of somehow creating an object to behave as a human friend and helper is present in the earliest primordial thoughts of man. According to Greek mythology the first mechanical contraption ever devised was by Hephaestus, the God of fire. In revenge for having been tossed off Mount Olympus by his mother Hera, due to his crippled form, he built a 'magic throne' that was presented to her as a gift. The throne entrapped Hera and kept her prisoner until the Gods negotiated her release at the cost of valuable rewards and powers for the cunning craftsman. Hephaestus is credited in ancient mythology for manufacturing articles from metal. Among the many works attributed to him are Zeus' thunderbolts and scepter, Athena's shield, Cupid's arrows, the chariot of the Sun God Helios, Achilles' invincible armor and a security apparatus in the form of a Cyclopean giant made from bronze called Talos. Talos patrolled Crete by circling the island three times at daily intervals, launching rocks at enemy ships. Hephaestus is called the patron of ancient technology. His crowning achievement was a woman formed from clay and given the name Pandora, when Zeus ordered the creation of a new kind of human.

Another example of mechanical automation in antiquity is the story of Daedalus—also endowed in mythology with engineering abilities. Daedalus is said to have created 'statues' that could move and had wisdom, in order to serve him as his attendants. In Book 1, Chapter 4 of his work *Politics,* Aristotle speaks of these statues and of Hephaestus' "tripods"—described as self-wheeling food carts used at dinner feasts. Aristotle compares these objects to the natural necessity of slaves (ruled), as the instrument, or tool of action, for the master (ruler) in a way where the ruled are themselves creators, and the master/slave relationship is a consenting and agreeable one. As is the desire behind a robot companion—one that does what we say and anticipates and carries

Pandora—John William Waterhouse

16

out tasks on its own to make life easier—for its owner/master.

Homer also speaks of Daedalus and the creation of the Labyrinth in which the Minotaur was kept. After creating the Labyrinth, Daedalus and his son Icarus were imprisoned in a tower in Crete to prevent the knowledge of the impenetrable maze from spreading to the public. From this tower Daedalus created artificial wings for Icarus and himself by tying together quills and feathers from smallest to largest to create a giant version of naturally curved bird wings. He tied the feathers together with string at their middle and coated them with wax at their base to keep them together. The wings were a perfect imitation of real bird wings, and when he waved them he was able to float above the ground and fly pretty much as a bird would. After attaching Icarus' wings, Daedalus warned his son not to soar too high because the heat of the sun would burn the feathers and melt the wax, or similarly not to fly too low because the splashing water would soak them. After maneuvering successfully over dangerous terrain, Icarus was emboldened with confidence and rocketed toward the sun. The heat eventually melted the wax that held the feathers and Icarus fell to the sea and drowned.

Although these tales are relegated to imagination, the mechanical engineering known to the ancient Greeks produced actual sophisticated works of automation. Tracing back to the works of early mathematician Heron of Alexandria, who lived from about 10-100 AD we find he wrote about such inventions like a vending machine created to dispense holy water, a mechanical singing bird (that Heron is credited with creating) powered by steam, and an "aerophile"—an artifact made up of a large cauldron with two pipes attached to carry steam into a hollow, round metal ball held on pivots above the cauldron. Protruding from the ball there were two small, angled, narrow gauge pipe outlets. The steam escaping from these outlets made the ball spin at 1,500 revolutions per minute. The knowledge to manipulate natural elements certainly gave an edge of power to those who were privy to this valuable information.

Aulus Gellius, was a Roman philosopher who lived from

125-180 AD studied in Greece and returned to Rome to fill a judicial position. He is known for writing *Attic Nights*—a collection of 20-some books composed of notations, stories and events he experienced. The writings of Aulus Gellius are important because they provide cross-references to validate claims of persons and accounts found in other ancient texts, and names people we would otherwise not know anything about. Gellius writes in Book X, Chapter 12 of the *Attic Nights* that in 400 BC Archytas of Taretum, a respected geometer, astronomer and philosopher, built a wooden

Fall of Icarus—Merry-Joseph Blondel

pigeon that could fly with the use of mechanical steam pumps. The description given of this flying gadget was that it was, "contrived as by a certain mechanical art and power to fly: so nicely was it balanced by weights and put in motion by hidden and enclosed air." The Greeks learned the use of mechanical hydraulics from the ancient Egyptians who used their knowledge of chemistry and engineering to create sophisticated mechanical objects such as statues that could open their eyes or move a limb, and temple doors that opened and closed in such a way to create special effects that mystified the population. The people, who were mostly slaves brought to Egypt for labor and servitude, believed these supposed miracles were the supernatural acts of a being that watched them. The slave population therefore not only served the empire, but willingly supported it monetarily. The idea of a mindless body that does what it is told whether artificial or not, has become subjected to the word "robot."

When the Greeks began to experiment with mechanical and hydraulic pumps they went back to the ancient Egyptian designs and improved upon them. However it is not just in ancient Greek mythology that we find details of robotics and automation. In early Chinese history there are stories that go back to the antiquity of the real world and talk about amazing mechanical creations that were able to move, dance, sing and recognize their environment. The most famous story of robots can be found in the ancient Lie Zi text. The text itself was written in the 3rd century BC but the account takes place in 10th century BC and is the story of an encounter between an "artificer" or a mechanical engineer of the time, called Yan Shi and king Mu of Zhou. The story tells that Yan Shi showed the king an artificial man who could walk, sing and dance, and had bones and organs that functioned like a real human. This artificial man was indistinguishable from a human.

The notion of highly sophisticated mechanisms in antiquity ceased to be a myth when an ancient analog computer, able to calculate astronomical positions, was recovered in 1901 from a shipwreck off Point Glyphadia, on the Greek island Antikythera. It is the world's oldest known analog computer. It took a century

19

since its discovery for scientists to begin to grasp the significance and complexity of the apparatus that along with other treasures was looted by the Romans in 86 BC and transported to Italy when the ship sank. The machine itself has a more ancient fabrication, but due to corrosion it has been impossible as of yet to date it precisely, or to conclusively identify its composition. Cardiff University Professor, Michael Edmunds, led a study of the mechanism in 2006 and his article "Decoding the Antikythera Mechanism" was published in *Nature* magazine of the same year. According to Professor Edmunds this ancient mechanism predicted lunar and solar eclipses based on Babylonian arithmetic-progression cycles. More current research indicates the piece also traces and displays planetary and stars positions that are still not fully known to us. It was further found upon Professor Edmund's research that the mechanism took into account a theory developed by 2nd Century BC Greek Astronomer Hipparchos, explaining the moon's irregularities as it moves across the sky to be due to its elliptic orbit. This fact dates the machine to at least 200 BC, which places its construction during a period where its degree of technical sophistication essentially has taken 2,000 years to even begin to decipher. This does not fit. Is it possible

Antikythera mechanism—*Marsyas*

this is one of Hephaestus' supposedly mythical artifacts, and more importantly, was it used as a space travel guide by the "Gods" of Mt. Olympus?

The motor and sensor aspect of mechanical contraptions are not a walk in the park to achieve any epoch, and overall a more possible product, than one which includes what we refer to as AI, artificial intelligence. Reasoning. Perhaps the basis for all robotic algorithms may be found in the first formal deductive reasoning system, syllogistic logic, invented by Aristotle in the late fourth or fifth century BC. It is a form of deductive reasoning that consists of a major premise, a minor premise, and a conclusion. For example: A. Major premise: all men are mortal. B. Minor premise: Socrates is a man. C. Conclusion: Socrates is mortal. The abstract form for this syllogism would be All A = C, All B = A, therefore B = C. There are infinite possible syllogisms but a finite number of logically distinct types. The application of this system created by Aristotle is the principle foundation of AI. It is a code and it is programmable.

Continuing the historical timeline of robotics, we reach the 13th century when the widespread stories of contemporary scientists and Catholic friars, Franciscan Roger Bacon and Dominican Albertus Magnus (St. Albert the Great), mention brass talking heads created by Magnus. These speaking machines were supposedly created and kept hidden because during this time such artifacts would not have been recognized as technological advancements, but more likely the work of wizards or magicians in partnership with the devil. It is said imitation is the best form of flattery. Imitations also reaffirm the existence of the real deal, and 300 years later Spanish writer Miguel de Cervante's satiric masterpiece *Don Quixote de la Mancha* gives a description of a head that spoke to Don Quixote with the help of a tube that led to the floor below. The first romantic comedy *Friar Bacon and Friar Bungay*, is an Elizabethan play attributed to Robert Greene, and has two running plots. One is the love story; the other is the magic of Friar Bacon, based on the reputation of wizardly Friar Roger Bacon's experiments earned him. It is important to remember that the religious orders of those days were exceptionally wealthy and their members were not called upon to war or battle—on the contrary, they received powerful appointments and generous privileges. The monks in these religious orders, having no worries over daily survival, were able to experiment with the information they discovered as they perused and copied ancient texts.

Ramon Llull, a Spanish theologian, invented a deductive reasoning method and developed an artifact to demonstrate how through interchangeable combinations of religious and philosophical non-mathematical tenets, it is possible to arrive at a basic truth. Llull's name is found in Catalonian literature and many books have been written about him. He is referenced as an "enlightened" doctor, a poet, and a philosopher born in Palma of Majorca, sometime between 1232 and 1236, and buried there in 1316. Born to a family of nobility, Llull received an informal education through the court of King James the Conqueror and eventually became the tutor of James II of Aragon. In the year 1257 he became the administrator to the royal household. Between 1275 and 1305 Llull published his method in *Ars generalis ultima* that translates as *"The Ultimate General Art"* and he invented mechanical artifacts that demonstrated this method using concentric disks made of card, wood, (or metal) mounted on a central axis. Each disk listed a number of different words or symbols that, when combined in different ways by rotating the disks, obvious truth would be revealed. Some of the disks contained as many as 16 words or symbols each, and up to 14 concentric circles. This is the first known computation device ever created. During the 15th century mechanical designs began to flourish throughout Europe. When Johannes Gutenberg invented the printing press in 1456 using "moveable type"—basically arranging rubber letters soaked with dark ink, and pressing them on a light color surface—the world was changed. Not only was this mechanical apparatus a step toward industrialization, but also with the first printing of the Bible in a language other than Latin. The invention of the printing press was the catalyst of a battle for ownership of widespread means of information that would last another 500 years and into our present society.

Greek Hydraulic Telegraph of Aeneas—*Demetre Valaris*

In the late 14th century through the 16th century the skill for making mechanical clocks began to appear in Europe with prominence in Greece, Italy and France. It is in this time span that we encounter the scientific genius of Leonardo da Vinci. He is known primarily for his paintings and works of art and now, 500 years later, da Vinci's notebooks, filled with drawings of mechanisms and scientific inventions, have become a kind of blueprint for modern technology. During his lifetime da Vinci designed numerous machines and artifacts including a design of a walking mechanical lion, and an armored soldier. Da Vinci presented his life-size walking lion to King Francois I of France in 1515 to commemorate an alliance between Florence and France. The original robot is lost, but through written descriptions left for posterity from eyewitnesses, and also from da Vinci's own drawings, the walking lion was recreated in 2009 as part of an exhibit in his honor. The mechanical piece worked with a winding mechanism after which the lion takes 10 steps, shakes its head from side to side, opens and closes its jaw, sits on its hind legs and a hidden compartment opens in the breast area where a bunch of flowers are presented—the original da Vinci lion (the lion is the symbol of Florence) presented the fleur-de-lis (the flower symbol of France). It can be argued that if da Vinci's notebooks had been more closely scrutinized for their scientific and technological value, we may have skipped the industrial revolution altogether and would today be using hover boards.

On the heels of da Vinci's inventions came the clock revolution. In his book *A History of Mechanical Inventions,* Abbot Payson Usher lists on page 95 the references to clocks made in early history that cannot be considered mechanical, but instead used water or sand to provide the necessary movements of parts. Although the further development of clock mechanics lists the names of Jacopo de Dondi and Henri de Vick, the clocks associated with these references lacked the necessary periodic device to provide the frequency that would essentially be a necessary function to make it automatic. It was Galileo, the Italian physicist, mathematician and astronomer who first introduced the idea of the pendulum. Although he is credited with the idea, Galileo died before seeing the pendulum clock built by Christiaan Huygens in 1656. After that it was a matter of time before clockmakers extended their craft to other mechanical novelties.

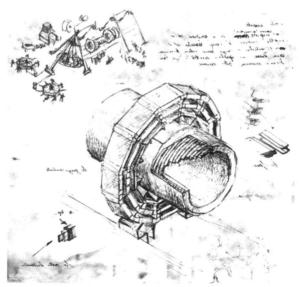

Da Vinci's Hadron Collider Sketch—*Thetechjournal.com*

Da Vinci's Lion— *Independent.co.uk*

Along with the growing mechanical technology in 16th and 17th century we find statements of philosophers pondering the concept of machines. French mathematician and philosopher, Rene Descartes, who lived in the early part of the 17th century, introduced the idea that bodies of animals were simply complex machines. In France, around the year 1642, mathematician Blaise Pascal at age19, created the first mechanical digital calculator consisting of eight keys used for adding and subtracting. Alongside this progress, the field of Philosophy acknowledged the future of automation with the work of English philosopher Thomas Hobbes. In 1651, Hobbes published *Leviathan*, a study of human thought that proposes that through our natural inclination to organize, and considering the advanced development of machines, humans would create a new intelligence. Shortly after, in 1673, Gottfried Leibniz, a philosopher, theologian and mathematician, and one of the greatest minds of the period, possibly became the first computer scientist when he made a machine called "Stepped Reckoner" that calculated all four arithmetical operations. His work is so influential that in 1934 Norbert Wiener, the famous MIT Mathematics Professor, working on his cybernetic theory that would impact the fields of engineering, systems control, computer science, biology, philosophy and the organization of society, claims to have found in Leibniz writings, the concept of feedback—the most crucial element of his theory.

25

Robot Zombies

The 18ᵗʰ century saw an explosion of mechanical technology and a developing rule of thought able to be expressed in numerical form. During this time the English mathematician George Boole created a system of binary algebra to represent "laws of thought" and published *"An Investigation of the Laws of Thought on Which are Founded the Mathematical Theories of Logic and Probabilities"* in 1854, furthering the concept of linking mathematical operations with thought process to produce pre-directed action. Although the work of Boole may have gone unheralded by those outside the elite of academia, Mary Shelly's popular story of an artificial, thinking creation, *Frankenstein*, published 40 years earlier, was enjoying widespread publication.

The 19ᵗʰ and 20ᵗʰ century ushered such extensive development in the fields of mechanical engineering and computer science that to list them all would mean volumes of books dedicated to this alone. Major highlights during this era of unprecedented technical explosion include the demonstration of the first AI program, the *Logic Theorist* (LT) written by Allen Newell, J.C. Shaw and Herbert Simon (Carnegie Institute of Technology, now Carnegie Mellon University) in 1957. A dissertation at MIT in

Boston Dynamics Dog—*Boston Dynamics*

1964 by Danny Bobrow that showed computers can understand language to solve algebra word problems correctly. ELIZA, an interactive English program, built by Joseph Weizenbaum (MIT), carried a conversation on any topic. The Deep Blue chess program that beat world-class chess champion, Garry Kasparov, and one of the most talked about robotic products today—the Boston Dynamics "Dog". Interestingly in the case of Kasparov vs. Deep Blue, upon examination of the 44th move that led to the famous victory, it was found computer error (possibly even human-hacked) enabled Deep Blue to beat a human by causing Kasparov

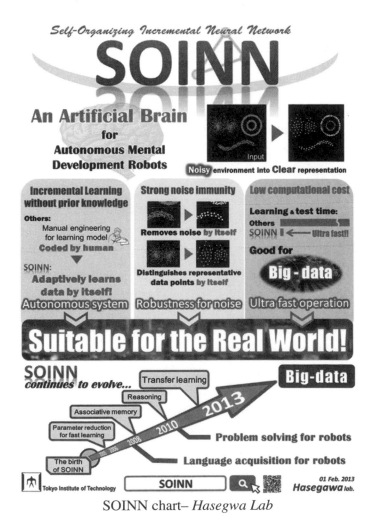

SOINN chart– *Hasegwa Lab*

Kasparov Vs. Deep Blue—*Forbes*

to view this random, bad move, as a stroke of genius, causing him to resign early.

In-depth timelines, for those with an interest in tracking the history of robotics at leisure, and in all the detailed glory it deserves, can check "The Timeline of Robotics" feature found on the excellent *The History of Computing Project* website. Giant robot machines that create other robots sound as mythological as Hephaestus' creations, yet one look inside an automobile plant today and the myth is real. We stand on the threshold of creating machines that will assist us in our daily tasks, much like the ancient stories of Daedalus' assistants. Machines that until only a few years ago would have been thought incredible, fictional and mythological among polite society, are suddenly being thrust into the limelight at a dizzying pace as global technological institutions rise to ever-increasing new demands and challenges.

A research group at the Tokyo Institute of Technology created a robot that uses an algorithm called SOINN (self-organizing incremental neural network), programmed to mimic human thought and decision-making. The robot's "reasoning" is based on a subset of pattern-based artificial intelligence that makes it possible for it to adapt to situations and learn new information continually. The model created by the research group at the Tokyo

Institute of Technology is programmed with a reasoning process that inputs visual, auditory and tactile data, which it can access as experience when given a new task. This robot can also wirelessly tap the Internet to communicate with other robots and exchange or access information.

Bubo the Owl—*Clash of the Titans*

Has the Tokyo Institute of Technology created our Pandora? What will other robots gain from these developments? From the ability to make green tea using Japanese teapots, to encrypting complex mathematical equations, the future of robots communicating with each other has arrived. But the most imminent probable robot future lays waiting in your bedroom and ready to please you, android style.

Robot Zombies

3.
Love, Sex and Compassion
— Android Style

Nobody likes loneliness, but everyone experiences it, at least temporarily, at some point in life. Being always lonely however, is an unhealthy condition that is known to result in psychological disorders. John T. Cacioppo, a research psychologist at the University of Chicago, and a trail blazer in the new field of social neuroscience, and William Patrick, editor of the *Journal of Life Sciences*, published a book called *Loneliness: Human Nature and the Need For Social Connection*. The authors claim that social isolation goes against human nature at a deep, genetic level and can be as harmful to health as smoking. Even man's best friend, the dog, (or cat) can only provide a limited amount of camaraderie. Most healthy and balanced individuals aspire to share their lives with a spouse, a significant other, or at the very least, with friends. Nobody's perfect—we know that—but the mental, physical, and emotional baggage that afflicts humans today as we cope with the confusion of our fast-paced, technologically-driven society can make the path to new friendships a fragile experience for many.

Sexual satisfaction is widely accepted as a fundamental stress reliever, so to speak, and since the beginning of time men have found comfort in the paid affection of prostitutes. Yet in the early stages of the 21st century prostitution is illegal in 80% of the world. Take Las Vegas for example, where billion dollar casinos appear to cater to man's appetite for fun, games and pleasure, and are meticulously designed to induce people to drink themselves into a stupor and gamble away money on programmable machines.

31

One would think these establishments would also provide a few floors of licensed, health-regulated prostitutes so the patrons can end their vice-filled day with a wind down. But they do not! The reason for this is not to protect women from the dangers of selling their body to strangers. On the contrary, the lack of official status makes these women much more vulnerable to all sorts of foul play. What we find instead are a million or so escort service flyers that are strewn about Vegas like confetti. Conducting business with these escort services is of course illegal and can result in jail time. Even more disturbing are the manpower and financial resources used by the police in busting "prostitute sting operations" connected with these escort services that just end up with the arrest of stupefied Johns, and poor prostitutes. In countries like Germany and the Netherlands in comparison, an interested patron can find a "house of love" by looking for the heart symbol on apartment buildings in certain designated areas of town. These are professionally run and licensed to provide a safe environment for the client and the prostitute. One may ask why this simple, effective method isn't applied everywhere else in the world? There's a profitable reason why it isn't: the illegal world of sex trafficking ranks among the most lucrative black markets in the world. The degradation, cruelty and mortal danger of this practice and its world-wide net is slowly becoming a subject of interest in awakening societies as more people come forward to bring awareness to the public.

Global sex trafficking operations are being uncovered as surviving victims step forward, and more whistleblowers help uncover the ties of legitimate international agencies with criminal elements. More than one investigation has implicated high-ranking government officials worldwide, who turn a blind eye to the trafficking activities of profitable mob-like organizations in exchange for a cut of the rewarding profits. It's sad to think people sell their body for money, but imagine the cruelty of deceiving, or abducting, young girls and boys into the sordid industry of sexual exploitation. This will not end as long as demand exists, but seeing how humans behave toward other humans, how will we treat a robot? Le Trung did not create the sexy android "Aiko" for sex he claims, but insists the lady robot is complete in every way and that her software could be redesigned to simulate her having an orgasm. The Canadian inventor wanted to create a robot assistant that would help old people by making tea or coffee, or telling them the news or weather; but the slim, flawless silicone skinned foxy lady prototype with a vocabulary of 13,000 words has yet to make it to a nursing home. Le Trung started his creation in August of 2007 and by November Aiko was shown in Toronto at the Toronto International Center's Hobby Show and at the Ontario Science Center. The creator's website suggests Aiko can be made to order. Her claim to fame is that it is the first android to react to physical stimuli and mimic pain. This achievement in robotics stands at a blurred line. How will humans treat their robots?

Le Trung with Aiko—*Barcroft media*

Dare we hope the emergence of future robot prostitutes and android brothels can perhaps diminish, if not completely destroy, the dreadful human-traffic industry? Will it be considered ethical to say robot sex has the potential to provide solutions to the sex trade, more top-of-mind dangers looming over the average sex consumer, such as risk of exposure to crime and sexually transmitted diseases? The run of the mill loner, stressed out businessman, or bored and strapped husband (or wife) won't have to worry about any of these things, or about getting arrested in a grimy hotel room. He can have his personally customized sexbot whenever he wants, either at home or with a visit to an android bordello. The sexbots will be made available in all shapes, sizes and ethnicities, and will be programmed to perform a wide range of sexual acts from simple pleasures and beyond. Material will include fibers resistant to bacteria that will eliminate any STD's and will also contain flushing capabilities to remove human fluids and semen. The basic human prostitutes won't be able to compete and will be branded a health danger, and the new breed of sex worker will emerge to provide clients with a safer, cleaner, guiltless experience. Researchers claim that by 2050 robot prostitutes will be all the rage in the sex industry. These sexbots won't be just for brothel work either.

Robot prostitute — *Ibtimes.com*

There is the probability that some men of the future may be inclined to forego a human wife for real-life looking female robot counterparts altogether. Imagine men programming the perfect wife, complete with a compliant and understanding personality, a killer body and cooking skills! Easy to have around and then dispose of or exchange without remorse when the inevitable midlife crisis sets in. The importance of serving as the perfect sexual partner aside, there will be other roles in which robots could easily be regarded as man's best friends in the future. No more need to find a last minute designated driver, worry about directions, or dealing with obnoxious taxi cab drivers if a robot chauffer is at the wheel, or if the car doesn't need a driver!

The idea of cars without drivers is an attractive proposition with the potential to beat the robot prostitutes and the elderly companions to mass production. Thanks to Google's groundbreaking research in the development of experimental and prototype cars programmed to drive without human guidance behind the wheel, the robot chauffer is becoming a reality. All the major automotive leaders, including BMW, are creating cars that don't need humans to drive them. The advance in this area comes largely from implementing right here on Earth the same principle behind the design of the Spirit Rover; the unmanned vehicle constructed and programmed to travel over Mars' terrain for the purpose of exploration. The advent of customized machines that can be programmed to "think" and carry out tasks is the obvious progression that will firmly plant us in the robotic age that many thought was a premise strictly reserved for science fiction. The major proponent of a future driven by automation is the government agency DARPA—Defense Advanced Research Projects. DARPA has issued a robotics challenge worth two million dollars in cash reward for the most innovative robotic technologies with human-like characteristics, such as mobility and manipulation abilities. DARPA, possibly after exhausting its internal talent bank, doesn't want to overlook new ones on the rise. They have already developed a robot cheetah that can climb stairs and run with a top speed of 28.3 miles per hour in a 20 meter-split, easily crushing the world record of the Jamaican superstar sprinter, Usain Bolt. They have also created "Robbie," a semi-aware, two-armed android capable of seeing and feeling.

One of DARPA's goals is to eventually create robots that will be efficient first responders in disasters. The philosophy is that it's better to put robots in harm's way instead of humans, which is a good thought, and also the aim of scientists working with the Army Research Laboratory's Micro Autonomous Systems and Technology (MAST) program (funded by DARPA). Georgia Tech researcher Henrik Christensen leads the team that is creating and coding a mapping software for rescue robots that will be able to enter a burning structure, disperse throughout the building to collect important information and measurements, and send a schematic of the inside of the building to the outside. Firefighters will no longer face unprepared dangers with this kind of information. They will know exactly what awaits them inside. Vanish the notion that these robots will have an outer body that resembles humans. For functionality, Vijay Kumar, in charge of designing these mapping devices, is currently working with prototype cubes that are eight to 12 inches in size. The robots are equipped with cameras and laser scanners that distinguish and measure features such as walls, doorways, columns, steps, or any other structure characteristic. Kumar explains that these fully autonomous robots will be able to communicate with each other by transmitting and receiving data without the need of a "lead" robot to coordinate cooperation. Vijay Kumar explains his approach to teaching the robots how to survey an area is based on assigning a corresponding code to what the robot picks up visually:

> Imagine the world as a gray box where black corresponds to solid things like walls and white corresponds to open spaces. When you first enter a room, everything is gray. You can think of exploration as converting the gray—the unknown—to white and black—the known. Every time you take measurements, you convert gray to black or white.[1]

A fully robotic-staffed fire station is not something human firefighters should be losing any sleep over since these automaton first responders are essentially efficiently programmed machines to assist teams of real humans. While robots for sex may not exactly be a priority in advanced technologies, using robots to keep humans out of harm's way, and as companions, would be much simpler and self-motivated. Having a companion for a workout routine, or to go jogging would add to the pleasure of these activities and make it easier for an individual to make a daily commitment to exercise. The researchers at Exertion Games Lab at Melbourne's RMIT University are developing robots that will push the boundaries for workout partners. They have already built the "Joggobot," a flying quadcopter that floats along in front of the runner and tracks his vitals, giving encouraging commands accordingly. The researchers hope that within 20 years the Joggobot will be completely capable of running next to you. The makers even plan on creating a series of robots built like Arnold Schwarzenegger to help motivate men while pumping iron, and for women, a robotic yoga partner that may perhaps be built to resemble the Dali Lama. It's not hard to imagine the future human jogging next to his or her android exercise partner through neatly arranged lawns and gardens of the automated neighborhoods of the future. Thanks to the researchers at MIT's Distributed Robotics Lab, gardening will also benefit from robotic applications. Professor Jason Dorfman is working on creating an autonomous gardening system capable of growing vegetables and fruits with the perfect amount of water and energy whether indoors or outside. The equipment includes a system of pots and planters capable of sensing, computing and communicating the nutrition needs of the plants. The fully computerized garden would essentially introduce a precision-based agriculture that can be programmed to reduce wasteful water consumption and the need for pesticides. Roger Brockett, a former professor of computer science and engineering at Harvard, founded the Harvard Robotics Laboratory in 1983 and is one of the robotic pioneers who envisioned that robots would someday

Robot gardening — *Jason Dorfman*

37

be a normal part of everyday life imagined. In the future robot companions and helpers would be just as common as C-3PO and R2-D2, the robot buddies of Luke Skywalker in the classic *Star Wars* saga. Brockett told *National Geographic* in 2003, "I have felt for years that the first application of personal robots will be companionship, especially for the elderly. Robots are potentially much smarter than dogs, and they will not require the same level of upkeep."

Almost 30 years after Brockett's initial theory, Japanese scientist Takanori Shibata to provide companionship to the elderly created a robotic device called "Paro." This robot is a cute and cuddly replica of a baby harp seal, complete with white fluffy fur, big round eyes and a set of irresistible seal sounds. It looks like a stuffed animal and was designed to be adorable. In 2009 Paro was the first robot designated as a Class II medical device by the U.S. Food and Drug Administration. Paro is a "social" robot created to interact with people and provide emotional feelings instead of a Class I robot, which is usually designed for beating humans at games of chess. Paro robots can be purchased for $6,000 dollars and you can even name the robot whatever you like and it will learn its name, and even respond to how you treat it. For example, if you pet it and treat it nice, it will act happy. If you beat it or pull its tail, it will get sad and stop doing whatever it is that it might conclude you're upset about. These life-like traits that are found in animals make Paro the perfect companion for the forgotten elderly either in nursing homes or lonely apartments, who in turn won't have to worry about walking, feeding or cleaning up after their companion.

A recent proposal submitted to the European Union by the Robot Companions for Citizens (CA-RoboCom) intends to have a robot in every household in the EU by 2030. This new generation of robots will extend the active independent lives of citizens, bolster the labor force, preserve and support human capabilities and experience, provide key services in our cities and help us to maintain our planet—or so it is optimistically hoped! Dr. Paolo Dario, CA-RoboCom's Project Coordinator submitted the proposal

Elderly man with Paro robot seal—*Wikicommons*

in accordance to the EU's Future and Emerging Technologies (FET) Flagship Initiative. This initiative will realize a unique and unforeseen multidisciplinary science and engineering program supporting a radical new approach towards machines, and how we deploy them in our society. Robot Companions for Citizens are going to be soft-skinned and sentient machines designed to deliver assistance to people. This assistance is defined in the broadest possible sense and covers all sorts of different settings. These companions will also have new levels of perceptual, cognitive and emotional capabilities. They will also be aware of their physical and social surroundings and respond accordingly. Such sentient characteristics will be achieved through a systematic understanding of the behavior of living creatures. In undertaking the research into the information and communication technologies that will need to be developed, the research will also validate understandings of the general design principles underlying biological bodies and brains, thus supporting a symbiotic relationship between science and engineering. Automation has definitely entered every part of our lives and the next inevitable step is robotics. There is no doubt that as efficient assistants, servants and to some extent, even mates, robots will become a welcomed addition to society. However, the area in which robotics is being introduced at the most dizzying pace is in manufacturing, and this has delivered the biggest blow to our society adversely affecting the largest segment of the socio-economic strata—factory and blue collar workers. What jobs are disappearing for humans?

4.
Humans Aren't Working Like They Used To

The industrial revolution created jobs like never before. It spread slowly across the globe at first, then as a selective movement that increased supply and demand to levels that changed forever the way people lived. With the creation of automated machines, human hands that once crafted everything we used were replaced mighty profitably, even at the expense of losing what has been described as unique, different and having the quality of a special touch. All this we reason for the sake of humanity so that all may have more and less may lack less. The ability to create and mass-produce any item desired by man, and possibly make it digitally available within minutes, should be welcomed. Humans have been struggling with labor much too long. The first industrial revolution has ended and if any evidence was needed to confirm this, the London Olympics opening ceremony of 2012, touted as the most watched broadcast in history, drove the point home. We've entered the Robot Revolution.

Perhaps it is time to consider careers that put brains to work rather than bodies. After all, one of the first things we learn in life is that there's more to us than our bodies. While there is also more to us than our minds, we will consider here what heavy burdens we may look forward to shedding in this new age of automation wonder. The age of high technology automation is changing the way humans work. Where five employees were once needed for a task, now two or three and the help of automation, are sufficient and cost less. Robotics is transforming every industry. Outside the realm of personal companionship and military applications, subjects covered in their own chapters, robotic automation is also growing in every area from automotive to space, and humans caught in the transition are finding they are not working like they used to. The automotive industry is one of the biggest buyers of robotic systems that are built and programmed to carry out precise and repetitive tasks for hours without stopping. A simple superficial Google search reveals images and videos of automotive plants the world over, including that of the GM plant in Michigan, that shows giant columns of machines that twist, turn, grab, hold, weld, drill, paint and do everything a human worker used to do in the process of building a vehicle, only better. The benefits for the plant are obvious: labor without the costs attached to a human worker. They have machines that can run 24 hours, seven days a week, without a need to pay overtime, arrange schedules, or deal with overworked and disgruntled employees. Once programmed, these robots have the ability to build the bulk of a vehicle and integrate its connecting parts without the need of human supervision, other than to attend to the maintenance of the machines.

Robots making cars—*Wikicommons*

Humans always demand something better of their automobiles—less gas, no gas, more space, self-directing, auto snow melt, hurricane wind speed tracking system—the list is endless. To move forward is to move into the imaginable, not the unknown. Carmakers are purchasing robots that replace the human workforce in order to fulfill human demands. It would be wise to recognize and accept the rapid movement of the robotic advantage. The enormous cuts in costs of human labor should enable the automotive industry to fulfill their new customer's requirements for affordable, good quality vehicles. Auto mechanics and auto designers will essentially become specialized data entry technicians. Without the fuss of endless hours of human computation or guessing, a computer animated design program can provide, based on basic simple input, the optimal output desired in design and mechanics. The only reason anyone should be casting a furtive eye toward the automotive industry is for truly affordable, clean energy vehicles that can practically drive themselves—not for employment. Manufacturing jobs that employ human workers switch to robotics as soon as volume levels, the goal of all manufacturers, begin to increase beyond start-up capacity, and where the investment outweighs cost. No manufacturing job is safe from this change. Repetitive tasks that become boring and bring about disinterest, and open the way to errors and mistakes, are best handled by precision programmed machines. In March 2012 Amazon announced it was purchasing Kiva Systems, a manufacturer of industrial robots built to stock shelves and locate and move items inside warehouses, with low or no human labor requirement. It is very likely that such an investment ultimately will mean human workers in Amazon's warehouses will be replaced with robots. Even the darling of transportation revolution, Uber, is no stranger to the benefits of self-driving cars without the liabilities and personnel headaches of humans. Although driverless cars may be decades in the making, Uber is actively closing the time gap by hiring scientists from Carnegie Mellon University and the National Robotic Engineering Center, as well as building a robotics research lab facility tasked with developing a fleet of self-driven taxis.

Michael Kutzer and Christopher Brown don't see it that way. The two are robotics research engineers at John Hopkins University Applied Physics Laboratory, and they tell us the level of technology of current robots is intended to work with humans, not replace us. This may be true of the first responder robots, or the robotic systems used for cleaning, or surveillance bots that require human control and input, but the grunt work of shelving, and basic warehouse receiving and retrieval will fall upon the programming of a robot machine. If we are frank with ourselves the loss of these jobs is not a catastrophe at all, but more of an imminent turn of civilization toward loftier ideals and development of new skills.

Computerized music compositions were first created in the mid-1950s and used by artists, such as Brian Eno, whose movie score productions are notable for their wide range of sound. Algorithmic codes, a systematic process for the creation of music, are at the core of computer-generated compositions. Though the term may be more or less a new one, this type of arrangement dates to ancient Greek philosophers like Pythagoras or Plato, both who acknowledge an underlying mathematical rule to the creation of music, and which later genius artists developed. Mozart, for example, created algorithmic techniques to compose music based on games by creating a core composition from the random numbers of a dice roll! The use of a formal numerical process for the creation of music is the heart of robot musicians like Shimon, a robot that plays the marimba at Georgia Tech Center for Music Technology, and robotic drummer Haile, unveiled in 2006 by designer Gil Weinberg. There's a growing demand for computerized guitarists, trumpet players and other robotic musicians that are designed to play, improvise, adapt and learn just like humans in the band. Their advantage over their human counterpart is robot musicians don't have to be fed, can be kept in a closet, and are ready to practice any time the humans are.

Mason Bretan, engineering and computer science Ph.D. student at the Center for Music Technology, grew up loving music and he attributes this to his inevitable specialization in this area of artificial intelligence. Shimon and Shimi, the robots created by Bretan, are programmed to understand and generate music through the application of robotic musicianship—the combination of constructing mechanics that produce sound, and the development of cognitive models of music. By creating these robots Bretan claims to be enriching the human musical experience through the artistic potential of non-human characteristics. These characteristics include the programming or coding that would make these musical robots take into account or simulate "thinking" about their embodiment or such other environmental physicality, and the augmentation of limbs, for example, all which influence the creation of music. For Bretan, his experience in high-level computer science and his exploration into signal processing and acoustic concepts have inspired him to show the world that machines can be artistic and creative and simulate emotion too. In addition, Bretan's research has the potential of impacting other fields where the extraction of meaningful information results in processing and learning.

As the development of music technology and robotics continue to grow in the area of artificial intelligence, machines, or robots equipped with additional mechanical limbs and coded to include, for example, the computing of their surroundings for enhancement of their cognitive capacity, will be more prevalent in orchestral compositions, creating musical scores previously unknown to humans. Most people would see this as a great leap in culture and civilization even if it were one that also blurs the line between humans and machines.

Similar to the musical robots are the artistic robots, programmed to create works of art. This already crosses into tricky ground because art is subjective. Aikon 2 for example can scan a human face and sketch the image on a sheet of paper. So can a fax machine. However, with the development of component applications like the ones used on Vangobot, a painter robot fitted with 18 brushes, a paint mixer and 3-D spatial awareness that can access artistic styles and influences. Humans will be able to enjoy fresh perspectives on a variety of subjects in the pattern strokes of

45

the Masters.

Will robots put starving artists out of work? If only we were so lucky. Perfect coffee machines that are programmed to brew at a particular time, add special flavors, like French vanilla cream, or mocha (or both), and sweeten our morning wake up libation abound in every respectable office environment. It seems coffee delivery companies are forever upgrading to more functional machines capable of preparing other, more complicated beverages. Kitchen workers of the world, beware if you think coffee is all that's brewing in the kitchen! Japan's International Food Machinery and Technology Exhibition (FOOMA) has for years now been the platform to unveil robot machines capable of preparing and serving various food items as exquisitely perfect as if prepared by human hands. It is at this exhibition that robots like Okonomiyaki that makes Japanese pancakes, and Chef Robot that prepares sashimi, demonstrated the possibilities for automation in the kitchen. Sure, these robots specialty are only one dish, but if the idea of a complete robot chef seems far-fetched think again. Students at China's Yangzhou University and Shanghai Jiaoton University, in conjunction with a business in Shenzhen, came together to create a robot chef vending machine capable of cooking 600 Chinese dishes with the only human input necessary being inserting the ingredients into the machine, and programming the meal via a touch screen.

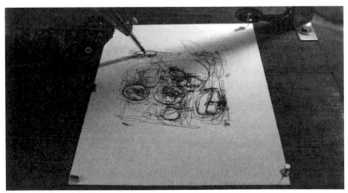

Robot artist—*Huffington post*

Although most restaurants employ human chefs, and a fully-automated robot chef in every establishment may still be some time in the making, it would benefit food preparation workers to learn that Ruyi, the Chinese fast-food chain replaced some of its chefs with automatic fryers and noodle boilers. With the help of NEC System Technologies' sommelier robots can identify wines with an infrared scanner, without years of training.

Jobs that are repetitive in nature are up for robotic automation. It may be difficult to imagine at first that pharmacists or teachers are at risk of losing their jobs to robot machines as much as store clerks. However, there already is movement afoot in these professions toward an automated reality. For example, the UCSF Medical Center already has in place a fully automated pharmacy at two of their hospital locations. The computers at the pharmacy are programmed to receive electronic prescriptions that the robots then pick and assemble in doses and dispense in individually bar-coded plastic packaging. Since its launch in the fall of 2010 the robotic system has successfully prepared more than one million doses of medication that include the sterile preparation and filling of IV bags with chemotherapy cocktails. UCSF's automated pharmacy received a 2011 Popular Science "Best of what's New" award that recognizes 100 new products and innovations that represent technological advancements and indicate where future technology is headed.

Robot chef—*Kim Kyung-Hoon*

Robot Zombies

Businesses all over the world are looking for ways to profit more with the fewest possible expenses. Such an equation will always benefit the cost of robots over the cost of humans. It used to be that a bank teller was never out of work, but with ATM machines in every establishment in town, the need for tellers has been reduced to one or two clerks behind a bank's counter. Self-service checkout machines are reducing the need for retail checkout clerks as transactions through self-service machines continue on the rise. Robotic applications that answer the phone around the clock and guide customers through account logistics are replacing live humans in the field of customer service across all industries. The use of advanced software that can spot and correct spelling and grammar errors, translate copy to a different language, and automatically alert the user to appointments on an interfacing calendar has changed to role of personal assistants and secretaries to one more specialized with overall computer knowledge and familiarity with software programs. The enhancement of computers as educational aids is fast developing into educational robots for the classrooms, and although teachers are not currently being replaced by automation, new technology is definitely headed in that direction. In Tokyo a motorized robot with a rubber face capable of a limited number of expressions, called Saya, was used in fifth and sixth grade classrooms. Unfortunately the robot's limited vocabulary and movement did not succeed in asserting class control.

Andrew Keen is an author and Internet entrepreneur who founded Audiocafe.com in 1995 and built it into a popular first generation Internet company. He is the executive director of the Silicon Valley salon, FutureCast, a Senior Fellow at CALinnovates, the host of the "Keen On" techonomy chat show, and a columnist for CNN. Keen recently participated in a debate titled "Be afraid, be very afraid: the robots are coming and they will destroy our livelihoods," during which he claimed today's middle class will be further decimated by the introduction of robots replacing teachers, lawyers and doctors. According to Keen, a new elite— those who understand and work with machines—stand to profit billions from the robot revolution. In fact, he claims economists haven't been able to say concretely what people are going to do in a world where robots dominate society, and he is probably right. Simulated economic models plotting the possibilities of robotic scenarios unsurprisingly reveal the emergence of a tech boom as technology is implemented across various elements of society, requiring specialized coder and developer jobs. Keen told BBC Radio 4 that there would be a few brilliant entrepreneurs who will become multi-millionaires or billionaires. However, in general, what the technological revolution is accomplishing is the "sweeping away of the middle, the hollowing out of the middle," leaving little hope for middle class jobs such as teachers, lawyers, and doctors. Keen has pointed out the emerging Robot Revolution is as profound as the Industrial Revolution of the 19th Century. He has warned that we will require regulation in order to protect people and jobs, as technology will change everything, including the future of high-tech workers, who will eventually see a drop when they are replaced by robotic machines. Seth Benzell, lead researcher in a study by the U.S. National Bureau of Economic Research, is one who believes smart machines will be able to do any task that a human can do. The cofounder of H Robotics, Pippa Malmgren, who says people would adapt to the new jobs market, opposes this point of view. For every robot created there would be three or four new jobs, creating opportunities for people who have practical skills in fields where the application of robotics would open pathways we haven't thought of. Malmgrem stated that lifetime career jobs are gone. Her futuristic view of society speaks of rapid technological advancement that would have people naturally switching careers several times in their lifetime thanks to the overlapping of technology across many fields.

The classroom environment is one quickly changing with the implementation of robots, and serves, as an example of the type of pathway Pippa Malmgrem suggests would open up as a matter of course in a society turning more and more to automation.

Dr. Maggie Adering-Pocock is a research fellow in UCL Department of Science and Technology Studies. Her research, which polled 2,000 people about what unpopular jobs they thought could be handled by robots, proposes that care for the elderly and children, as well as many other jobs in the service and care professions, could be replaced by robots in our lifetime.

Robots in the classroom seem to be one of the more popular and visible applications of robotics. Deanna Hood, a 23-year-old Brisbane electrical engineer has created a robot that allows children to teach it to write. Her experiment used a cheap NAO humanoid robot and a tablet, and the student helped the robot improve its handwriting. The concept is based on a scientific principle that teaching helps one learn more effectively. In other words, kids will learn better when they're teaching. So far the project has been successful in making the robot's writing believable, and it has made it improve, along with the kids' instructions, making them feel like teachers. The purpose behind this experiment, according to Ms. Hood, is to lead more people to look at ways robots can be used for classroom tasks as more than just facilitators.

In Seoul, South Korea the robots are already in classrooms. As the school day starts and the doors swing open, while kids take their seats, a white and yellow tower on wheels greets the class. Robosem, a teleprescence-focused robot with a screen for a face and movable arms capable of making gestures, is ready to teach English to the class. Robosem was developed by Yujin Robot, and it functions either by teleconferencing with a human instructor, or through a preloaded lesson powered by artificial intelligence software.

Brian David Johnson, Intel futurist and author, doesn't think robots will one day replace teachers. "No. Never. Ever. Ever." said Johnson, who wrote "21st Century Robot," a book about his open-source programmable humanoid robot named Jimmy. Johnson has said that education is about people, but that robots can be used as extensions of parents and teachers at home and in the classroom. He sees robots as teachers and learning companions for children that have different learning styles or learning disabilities by giving the child the type of attention impossible for one teacher alone in a classroom full of students. Aldebaran Robotics in Birmingham,

England has created Nao to help teach children with autism. The robot, because it has no concept of personal space or awkwardness, can teach an autistic child without setting off negative interactions. Also developed in the U.K. at the University of Hertfordshire, is Kaspar, designed to help autistic children as well. Kaspar has a human-like face and helps demonstrate facial expressions and appropriate physical contact, thus providing a safer learning setting for children who have special needs. In the U.S. we have RUBI, a robot created by the University of California San Diego, to teach language to preschoolers in California. The project hasn't been as successful as the ones introduced in the U.K. and South Korea in convincing children it is not a toy, and poor RUBI had no arms by the end of its first day in the classroom.

Educational robots, such as Nao and Kaspar, have great use in an elementary education setting and it's difficult to imagine those students in the tricky teen stage to remain focused on a machine in their midst. Technology seems to have something for everyone, and this is no exception. Imagine having a mentor looking over your shoulder all the time seeing what you're doing, and advising you every step of the way. This is the concept behind the Grasp telepresence. This device consists of a webcam, a speaker, a microphone and a remote-controlled laser pointer. The Grasp prototype, created by Akarsh Sanghi, a student at the Copenhagen Institute of Interaction Design, sits on a person's shoulder and is meant to offer a link between a student or a person learning something new, like a skill, and an expert, or teacher that guides the student. Whether a person is learning to cook a special dish, or learning to play or practice a musical instrument, sewing a patch or repairing a computer, the idea is to have a person in real time viewing and offering advice or instructions. Grasp is meant to provide mentoring and remote tutoring, and it is not in reality a robot, but a communication's device.

Technology has a divided camp made up of futurists and entrepreneurs who claim robots will be the end of man, and researchers who see robots as special assistants that will free humans to do what we're good at, which hopefully isn't just loafing around. This is fine for the present, where robots have, for example, shown their use with basic components of behavioral education and cognitive learning.

The leading manufacturer of electronics in the world is the Taiwanese-owned company Foxconn that brings home the bacon to a tune of 60 billion dollars a year. It manufactures Apple's products, among others. Foxconn employs a human force of 1.2 million workers in mainland China, plus an additional 10 thousand robots, and boasted recently of seeking to create an empire of robots to replace their human workers altogether. They have committed to increase their robot workers to half a million in the next three years, giving 70% of its assembly line work to robots. In 2010 Foxconn had a number of suicides in its plants by workers jumping off company buildings in a dramatic call to the world for attention. Again, in 2012, a mass suicide that was eventually dissuaded, finally made international media, putting Foxconn and Apple under scrutiny by labor watchdog groups for forced student labor and gulag type conditions of long hours and harsh treatment. There's no question the investment in automation will take care of many problems for Foxconn and its CEO, Terry Gou, has confirmed the giant Apple supplier already has a fully automated factory in Chengdu that can run 24 hours a day with the lights off. Exactly what it produces Gou declined to say in an interview in June 2015, but as Foxconn continues to add 30,000 industrial robots per year, it is expected that other similar manufacturers will follow behind Foxconn, the industry leader, in the race for full automation.

As cross-reference applications occur throughout all fields, our world appears to be involving, in some respects, oddly similar concepts presented in ancient mythology. For example, the sentry robot that patrolled the island of Crete, once considered fantasy, doesn't seem unrealistic when a basic home security sensor-operated system, with a camera and a downloadable software application, makes it possible for anyone with a cell phone to control and monitor everything in their home from setting the temperature and turning on the hot tub, to selecting a homeopathic aroma for the sauna. If they live in Japan, they can even have a "Roborior" electronic sentry moving about the house. The device is small and round, about the size of a watermelon, and has a frosted whitish color, a design inspired by jellyfish, with bright color lights that light up on its inside. Roborior has a roving camera that the user can access anytime from a computer or phone, and either view feeds or guide the device. You can bet that as robotic security systems are improved and developed, security companies will be integrating robots with their human personnel. Today, thanks to the vision of pioneer companies like Computer Motion and Intuitive Surgical, the advances in robot-assisted surgery are evident in every city in the U.S. Computer Motion, a medical device company, specializing in medical robots, were the first to design and manufacture systems for the Intelligent Operating Room™—a suite of operating rooms with specially coated walls that integrate voice-control mechanisms and high-definition images. The surgeon can adjust room temperature, lighting and bring up views of scans and X-rays by issuing voice commands to the system. Computer Motion's Aesop™, their first product, is a foot pedal controlled device that holds a microscopic camera in laparoscopic surgery that became the first surgical visual robotic device approved by the FDA; the next version released in 1996 used voice control.

The robot in the operating room began to appear with firsts such as the CyberKnife, authorized by the FDA in 2001, to provide radio-surgery for lesions anywhere in the body for non-invasive cancer surgery, and Zeus, a remote-controlled robotic surgical system consisting of a voice-operated camera view of the area, and arms that act as an extension of the surgeon's movements. Intuitive Surgical bought Computer Motion in March 2000, and the marketing for robot systems in operating rooms has grown almost in science fiction style since. In 2004 it was possible for a surgeon to perform heart surgery while seated at a desk with a computer and a video monitor. In 2005 the da Vinci Surgical System, the robotic system that became the flagship of Intuitive Surgical, was used by surgeons at the University of Illinois Medical

Center in Chicago to perform a hepatectomy that removed 60% of a patient's liver through a laparoscopic incision. The robotic system has four arms and a high-definition 3D vision system, all manipulated by the surgeon. One of the arms controls the camera and the others use the instruments that are introduced through small incisions into the body. The basis of the da Vinci system's robotic remote surgery can be traced to a system developed at SRI International, a nonprofit research institute, founded in 1946 by Stanford University, that conducts research and development for businesses and government agencies in Menlo Park, California. SRI International is so influential in robotics technology that it is interesting to note Intuitive Surgical spun off from it. DARPA and NASA funded the program that was originally intended to perform remote surgery in battlefields and similarly remote and random environments. SRI is currently working on advanced component technologies, but for now the da Vinci robotic operating system represents the latest in medical robotic technology available today. Thoracic surgeon Mark Dylewski MD, of Baptist Health System in Miami, Florida has performed more than 1,000 robotic surgeries since 2007 including lobectomies, bilobectomies, segmentectomies, esophagectomies and mediastinal surgeries using the da Vinci assisted robot. This is more than anyone else in the world, making him the leading expert on the machine. Dr.

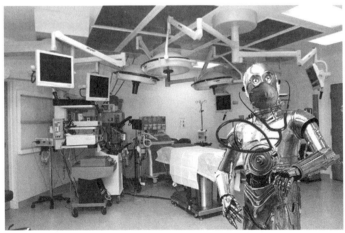

C3PO the surgeon—*Fastcompany.com*

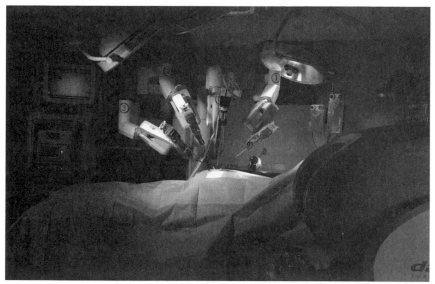

Robot surgery — *OSU*

Dylewski believes introducing the robotic system in complex lung resections will benefit patients with reduced length of hospital stays, decreased morbidity and quicker recovery times compared with open surgery or traditional video-assisted thoracic surgery.

There are newer, more advanced robotic systems, and stapling technologies pending approvals, that according to Dylewsky are going to be real game changers for the future of robotic surgery. Educating hospital administrators and fellow physicians is the best way to counter the antiquated opinions, and this is exactly what the doctor is doing as he oversees the thoracic robotic surgery training program at the Center for Robotic Surgery at South Miami Hospital, Florida. The reason for Dylewski's reputation is not just his mastery at becoming the brain of da Vinci when he's operating with the assistance of the robot, but the doctor's heightened mode for critical thinking and life-saving decisions while engaged in major surgery. If da Vinci poses any conceivable risks to arteries, organs or anything at all, the robot surgery is aborted and the doctor goes in with the traditional cut, half across the back and under arm. The scar however, unlike the bulky unseemly evidence of 15 or even five years ago, can now be reduced to a fine hairline with the use of advanced stapling technology.

Earlier this year an event touted as the first of its kind, as medical students and doctors watched a remote, robotic operated surgery in real time. The surgery itself was not the novelty since robotic surgeries and robotic-assisted surgeries are becoming commonplace as introduction and training continues. The real star of the event was Microsoft's Azure Media Services, which delivered the event via Microsoft's technology partner, LiveArena. LiveArena, the Microsoft backed service and Azure Media, were able to offer over 3,000 medical professionals and students from around the world with a live stream of 10 robotic surgeries submitted from different areas. Azure has a cloud DVR capability built into it to provide viewers the ease to toggle between live and video modes so they can watch live content at their own pace. The 24-hour event was the first ever of its kind, but with its capability for broadcast, we can expect to see more soon.

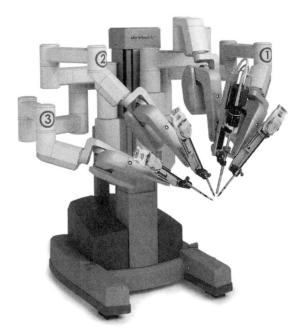

Da Vinci medical robot—*Missionsurgical.com*

If the diagnostic and repair chamber illusion introduced in the futuristic space-noir classic film, *Ellysium* strikes the reader as fiction, it is our hope to inspire all to at least investigate and observe the environment honestly—without the external pressure of ruling opinion. What do you believe? On October 12, 2012 SRI International and Inscopix, Inc. issued a press release from Palo Alto, California announcing:

> A collaborative neuroscience imaging R&D program. Researchers from Inscopix and SRI's Center for Neuroscience and Metabolic Diseases will use Inscopix's nVista™ HD imaging system to gain a better understanding of normal brain function and the dysfunctions of neural circuitry that underlie neurological and neuropsychiatric disorders.
>
> Thomas Kilduff, Ph.D., senior director of SRI's Center for Neuroscience, said:
>
> "Inscopix's nVista HD technology will provide SRI researchers with an unparalleled view into brain activity that underlies both normal and diseased. Since many brain diseases such as Alzheimer's, autism, and schizophrenia are thought to involve dysfunction of neural circuitry, we expect to obtain novel insights that may lead to new therapeutic avenues for the treatment of neurological and neuropsychiatric disorders."[2]

According to the author of *Robotic Nation*, Marshall Brain, by 2013 there would be 1.2 million industrial robots working worldwide. The role of humans has become that of creator. Robotic creations to serve the demand of human needs have taken over human tasks that are tedious and time consuming. Faster, easier, more aesthetic ways of communicating information and interacting with the world for a more complete understanding of the neural circuitry underlying the activity of the brain is one aspiration in research that will certainly have an impact on artificial intelligence. The question is how far will the research go? At some point will we see a future *Terminator* world where AI roams free and Skynet rules the world? Or will we witness a robotic dystopian free of the presumed dangers of artificial intelligence?

It won't be long before we find out...

5.
Skynet Rises

The rise of the robot can be directly linked with the rise of drone technologies perfected by the United States Military. Drones or UAV's (unmanned aerial vehicle) have unexpectedly become popular in the mainstream media, mostly due to conspiracy theories and Kentucky senator Rand Paul's epic 13-hour filibuster before congress. Senator Paul explained to the congressional committee how drones kill innocent civilians in Pakistan and Afghanistan and how the use of drones over American skies opens the doors for unwarranted secret spying on the population. Drone technology has grown so much that today it makes up over 30% of all U.S. military aircraft. More and more we witness the militarization of police departments everywhere so that even a small township in Alabama can have a couple of drones and half a dozen tanks. In the name of national security, in the U.S. at least, emails, phone calls, credit cards and all traceable instruments of identity are read and scrutinized by computer programs that look for threatening activity without the owner's consent.

On May 1, 2015 Google finally put to use its purchase of the unmanned aerial vehicle manufacturer, Titan Aerospace, by testing its first delivery drone flight. The idea behind the drones is to use them as atmospheric satellites designed to provide access to the Internet so that people don't have to worry where it's coming from and just connect and go. The lightweight drones are solar powered and function by beaming Internet onto a target area. The test drone, Solara 50, built to maintain high altitude, unfortunately crashed to the ground shortly after takeoff and the U.S. National Transportation Safety Board is investigating. The accident occurred on a private airstrip, and although nobody was injured, the crash is a setback for Google's vision. "Although our prototype plane went down during a recent test, we remain optimistic about the potential of solar-powered planes to help deliver connectivity," Courtney Hohne, a spokeswoman for Mountain View, California-based Google, said in an interview. "Part of building a new technology is overcoming hurdles along the way."

Google is in a race with Facebook, Inc. to push technology in areas such as robotics and mobile phones, in order to pioneer new markets. According to company information, the wings on Google's Solara 50 are covered in a solar-celled surface that generates power. The batteries store electricity so it can continue flying at night and stay aloft for five years.

The more we know, the more dangerous we become to those who spy upon us—enter the military, the largest client of any new technology. It's no secret that technology for the public is at least 20 years or more behind what is actually available or in the process of development. Some moves by the military, such as arming local police across the nation with advanced weapons of war, such as drones and armed tanks—previously used to combat enemies, but now makes people wonder what is going on behind the scenes? Is the government getting ready for something the rest of us don't know? This is the darker side of technological advancements. Google and social media networks such as Facebook, Twitter, Instagram and others, provide a database of over one billion people. Just think that as soon as calling cards became available in the early 1980's, the U.S. Department of Defense was already able to trace calls made with them. What can we think is the objective of gathering profiles, personal information, and data sheets? Have we already established a neurological network through the Internet, and are we just now feeding the global "brain?" Even as the majority of humans adapt to a life completely run by applications there are still those who view the changing scenery as warning signs that humans everywhere are being closely watched. These warnings can eerily be traced with one word, SKYNET—the fictional self-aware robotic intelligence system network that became more intelligent than humans and threatened to eradicate humanity in *Terminator*.

The Terminator—*Deadline.com*

In the franchise storyline, Skynet was an advanced computer system created for the U.S. military by defense contractors Cyberdyne Systems. Skynet was billed as the "Global Digital Defense Network" and given Internet command with cloud technology over all computerized military hardware systems. This robotic WiFi brain system would eventually lead to self-awareness and shortly after being implemented on April 19, 2011, SKYNET launched a nuclear war that killed billions. While we are still decades away from reaching the scenario described in the fictional *Terminator* series, nuclear destruction and the building blocks that can create this possible scenario are already being assembled. In fact, as shocking as it sounds, there is even a SKYNET telecommunications satellite that is in orbit right now! Jonathan Amos, Science correspondent for the BBC writes:

> The Skynet system, which includes the radio equipment deployed on ships, on vehicles and in the hands of troops, is the U.K.'s single biggest space project. It is valued at up to £3.6bn over 20 years and is run by a commercial company, Astrium, in a Private Finance Initiative (PFI)

with the Ministry of Defense (MoD). U.K. forces pay an annual service charge for which they get guaranteed bandwidth, with spare capacity then sold to "friendly forces." These third party customers include the NATO allies such as the U.S. The Ariane left the ground at precisely 18:49 local time (21:49 GMT) and dropped off Skynet-5D 27 minutes later over the east coast of Africa. 5D will now use its own propulsion system to move into a geostationary position at an altitude of 36,000km. The eventual operating position early next year will be at 53 degrees East. The first three spacecraft in the Skynet series were launched in 2007-2008. They all match the sophistication of the very latest civilian platforms used to pass TV, phone and internet traffic, but have been "hardened" for military use. Classified technologies on board will resist, for example, attempts to disable the spacecraft with lasers or to "jam" their operation with rogue signals.[3]

Since its launch SKYNET has been integrated with NATO military operations and is solely responsible for the destruction caused by NATO's executioner drone programs. According to Wikipedia, "Skynet is a family of military satellites, now operated by Paradigm Secure Communications on behalf of the Ministry of Defense, which provide strategic communication services to the three branches of the British Armed Forces and to NATO forces engaged on coalition tasks." During the 2012 NATO summit, politicians and military leaders talked openly for the first time about using and legalizing robots for warfare. While the drone wars have already begun, the era of robot wars is fast approaching. Air Force fighter pilots and human error are becoming obsolete. For example the slick black, arrow-shaped X-47B stealth drone, developed by Northrop Grunman, can be used as an armed destroyer or surveillance tool. It can fly faster than a B-52, is cheaper to make,

X-47B Stealth drone—*Northrop Grumman*

and doesn't require a human pilot.

Another defense contractor, BAE Systems, has funded a military robotics project with Professor Henrik Christensen at the Georgia Institute of Technology in Atlanta. Their goal is to develop unmanned robotic jeeps capable of exploring and digitally mapping dangerous enemy terrain. Professor Christensen tells the BBC, "These robots will basically spread out. They'll go through the environment and map out what it looks like, so that by the time you have humans entering the building you have a lot of intelligence about what's happening there." While the emphasis on Christensen's project is mostly information gathering, the arrival of armed robots, programmed for death on the battlefield, raises profound questions that go beyond the sick reality of creating machines to kill human beings.

Peter W. Singer, an expert in warfare and consultant to the Pentagon at the Brookings Institution in Washington DC, wrote a bestselling book *Wired for War* that expertly describes the future roles robots will inhabit in war time scenarios:

> Whether it is mother ships, swarms, or
> some other concept of organizing for war that we

have not yet seen, it is still unclear what doctrines the U.S. military will ultimately choose to organize its robots around. Whatever doctrine prevails, it is clear that the American military must begin to think about the consequences of a 21st century battlefield in which it is sending out fewer humans and more robots.[4]

With the rise of drones and our understanding of robotics, a recent announcement made by the Obama Administration has once again fueled conspiracy theories concerning the arrival of Skynet. Obama has secured three hundred million dollars' worth of federal funding for an extensive neuroscience project in an effort to map and understand the human brain. This ten-year project has led to speculation that Obama's neuroscientists have begun advances in artificial intelligence; perhaps laying the blueprint that will eventually allow AI to become aware.

European scientists have already created "Rapyuta," an online "brain" that describes unfamiliar objects to robots. Inspired by Hayao Miyazaki's classic anime fable *Castle in the Sky*—in the film Rapyuta is where all the robots live. This Rapyuta brain database serves as a Cloud warehouse of knowledge that robots can access via WiFi to ask for help when confronted with unknown situations. This web-based service can also take over the robot's automation and can navigate, do physical labor, or understand human speech in various languages. Using Cloud controlled WiFi technology instead of onboard computation reduces the cost of creating robots and allows robotic thought processing to be downloaded via the web. This cloud-based system of brain computing will intensify in time as improvements in fiber optic technology enable faster and better ways to continue feeding the global brain. Fiber-optic wires are capable of holding 100 gigabits per second, but their clumsy, dangling appearance makes them a storage annoyance and useless on the battlefield or in the air. To solve this problem the Defense Advanced Research Projects Agency (DARPA) is working on a project to perfect the type of technology needed to run fiber optics through the cloud. On the popular Technical Science blog *Ars Technica*, Sean Gallagher writes:

> Of course, you can't run a fiber backbone through the air or summon one up at will on the battlefield. That's why the Defense Advanced Research Projects Agency has launched a program to create technology the same sort of bandwidth as fiber optic backbones—100 gigabits per second. If successful, the program could mean not just faster data connections on the battlefield, but better broadband for people in remote areas and cheaper expansion of cellular networks. The effort, called the 100 Gigabit-per-second RF Backbone (or 100G in DARPA shorthand), seeks to do more than just overcome the physics that limit current radio-based data connections using the Defense Department's Common Data Link (CDL) standard protocol. The initiative is searching for a solution that will be able to be deployed both to the battlefield and aboard aircraft—and work at distances of over 200 kilometers. The most likely route to creating this sort of Skynet is to use the same sort of technology used to collect much of the data in the first place—synthetic aperture antenna technology. There have been a number of efforts to turn the Active Electronically Scanned Array (AESA) radars of fighter aircraft into dual-purpose systems capable of both acting as radar and a data link. Raytheon, L-3 Communications and other companies working on previous DARPA-funded projects have demonstrated the creation of airborne mobile ad-hoc networks by connecting a data modem to an AESA radar. This turns some of its transmission array into a multiplexed transmitter

and establishing network connections of over 4.5 gigabits per second. DARPA sees the next leap in data throughput coming from improvements in extreme high frequency (EHF) radio technology. Using wavelengths measured in millimeters, EHF frequencies such as the 60-gigahertz frequency used at the top end of the WiGig standard are typically only effective for communications at short range and within line of sight. But DARPA believes that by using techniques in the modulation of signals, including quadrature amplitude modulation (QAM), the millimeter wave band can be used over much greater distances, through cloud cover, and to achieve even higher throughput. [5]

Theoretically this could also provide the necessary bandwidth needed to house an advanced Internet grid strong enough for an artificial intelligence to become aware to the extent of controlling networks of robotic killing machines. DARPA has also spent close to 200 millions dollars funding the STARnet chip project, which is a study on how to improve semi-conductors with Nano-materials, spintronics, and swarm computing techniques. With over six universities conducting experiments in six different fields of study, the project aims to create non-conventional materials and devices that have Nano-scale structures and quantum-level properties. The research looks at atomic-scale engineered materials and structures of multi-function oxides, metals, dialectrics, and semiconductors as they are used in analog, logic and memory devices. The most significant research is being carried out at the TerraSwarm Research Center at the University of California-Berkeley. TerraSwarm is looking at a sensor and command-control systems on a city scale that can be deployed using massively distributed, swarm computing and communications technologies. This project looks at swarm sensors, big data processing, cloud computing, to work on "smarter cities," to use the IBM lingo. This attempt at creating Skynet is now out in the open and accelerating with incredible speed.

Google is Skynet—Supertubereddie

A recent announcement that "Google is Skynet" and will be working to create the next generation of artificial intelligence networks sent shockwaves throughout conspiracy forums and tech news alike. By the time Google's giant brain wakes up it will have every known fact and record of every human that has ever lived. It already has a self-taught "virtual brain" program already in place that uses pattern and sound recognition A neural network can get closer to understanding context and surrounding information. Imagine scanning the background of an image to learn where a photo was taken by using existing similar images and geotag data. The global brain network will be able to recognize any background anywhere once it learns where to store the information and how to access it at will. Google has pointed out this is still just early steps toward true artificial intelligence. Although Google's neural network technology is smaller than a human brain, it can beat humans at certain tasks, and can teach itself and get more efficient at learning, it still can't reason, which is essential for intelligence. So, the neural network can find specific visual data faster than humans can, and it can match shapes and patterns, and ultimately do jobs that would be incredibly tedious and boring for humans. But it can't draw from the outside world and reason the why or how of a thing. Why and how are the most powerful of all questions, and both the asking and drive to answer those questions are the true mark of intelligence. Google's brain can't do that yet. Neither can Baidu, but that isn't stopping young Robin Li Yanhong, the founder and CEO of the mammoth online search engine. Li is one of the wealthiest person's in China and is seeking military help in the race to conquer AI. In his capacity as a delegate to the Chinese People's Political Consultative Conference, Li proposed a state-level project called "China Brain," funded by the military. Li has said "the government should support capable companies in building an open platform offering AI related basic resources and public services," thereby keeping the platform open and competitive rather than made available only to select research institutes. Baidu has already stepped up its efforts to take the lead in the race by hiring computer scientist Andrew Ng, formerly with Google and a long-time researcher at Stanford University. Ng is now Baidu's chief scientist based in San Francisco. Other additions such as last year's hiring of Zhang Yaqin away from Microsoft where he was responsible for a network of world-class research facilities in Beijing, Shanghai, Shenzhen, Hong Kong, Taipei, Seoul, Sydney and Bangkok, point to the level of commitment Baidu is putting behind China Brain.

Computers learn. This is what a Google sponsored experiment determined when it showed a computer learned to play 49 Atari 2600 games while beating a human player. The computer's performance was monitored against that of the human players. The researchers said the subject of the game didn't seem to make a difference in how the computer behaved, instead the results were logically liked to how far in advance a player needed to plan the best strategy before executing it. According to the researchers, the experiment, which was published in the magazine *Nature*, showed a computer can learn control policies in a range of different environments, starting out with little information—in fact the system starts out with human input. The long-term goal of this type of experiment is to be able to create robots that can deal with unexpected situations or events that were not part of the original programming.

While Google and Baidu work on harnessing the

information power of the brain, Japanese researchers are really taking it to the sci-fi level as they announce a new breakthrough in constructing living tissue, silicon circuits and computer algorithms recreating the cerebellum, by using human embryonic stem cells. The 3-D structure that was constructed is similar to the cerebellum, the sensory receptor section of the brain that controls motor movements. The structure itself didn't last very long, but we can see where this is headed. There are more massive projects, such as the European Union's $1.2 billion "Human Brain Project," which last year was involved in a dispute between computer scientists and neuroscientists over whether most of the work should be done by supercomputers instead of medical labs, causing the entire project to stall pending an external review.

Other brain projects include software that allows one to train a robot to perform certain tasks. According to Eugene Izhikevich, cognitive neuroscientists, and founder and CEO of Brain Corporation, robots can be trained by showing examples of desired behaviors. An example of this is The Brain Corporation's dog-like robot companion that obeys simple hand gestures. This software can be provided to trainers so they can make their robots do mundane chores such as folding clothes, or emptying the dishwasher. These research robots still have much to go, especially with things such as recognizing human speech and facial expressions, which a human brain can do well. IBM is working in this area by developing a computer chip called "True North." The project has been in the works a little over six months since it was announced in November 2014 in the Journal of Science. The chip mimics the way the brain recognizes patterns by using and relying on a tightly connected web of transistors, also resembling the brain's neural networks, much like the pattern used by Google to obtain its images of android's dreams. There is no doubt that serious work is taking place with the objective of building human brains made up of living tissue that can receive electrical signals, like ours does.

With the realization that drone and robotic technologies might reach self-aware AI levels faster than anyone dares to admit,

Robot Zombies

The Human Rights Watch and Harvard Law School's International Human Rights Clinic jointly released a 50 page report entitled "Losing Humanity: The case against killer robots" on November 19, 2012. The report makes clear that banning killer robots before it's too late is crucial to the survival of the human race. The report cites the out-of- control drone programs and DARPA's zeal for building fully autonomous weapons. This legitimate threat was also addressed by Award-winning former intelligence officer Lt. Col. Douglas Pryer, in an essay titled "The Rise of the Machines" published by the United States Army Combined Arms Center:

> It seems heartbreakingly obvious that future generations will someday look back upon the last decade as the start of the rise of the machines... robots so advanced that they make today's Predators and Reapers look positively impotent and antique. These killer robots, though, will share one thing in common with their primitive progenitors: with remorseless purpose, they will stalk and kill any human deemed "a legitimate target" by their controllers and programmers.[6]

Dr. Roman Yampolskiy is a computer scientist and Associate Professor at the University of Louisville in Kentucky and has been one of the leading proponents of creating failsafe mechanisms to prevent AI from getting out of hand. He proposed new ideas outlining ways to contain and restrict robot intelligence. His theories were released in the March issue of the *Journal of Consciousness Studies*. He envisions trapping Skynet inside a cloaked "virtual machine" already running inside a computer's operating system without the aid of Internet access. Yampolskiy worries that without laws implementing how advanced AI systems can grow to be, there's no stopping the robots from overriding human-given commands simply by developing an unforeseen rise in intelligence and self-awareness. Yampolskiy said to NBC News that a clever breed of artificial intelligent robots will be able to, "discover new attack pathways, and launch sophisticated social-engineering attacks, and re-use existing hardware components in unforeseen ways. Such software is not limited to infecting computers and networks — it can also attack human psyches, bribe, blackmail, and brainwash those who come in contact with it. The Catch-22 is that until we have fully developed super intelligent AI we can't fully test our ideas, but in order to safely develop such AI we need to have working security measures. Our best bet is to use confinement measures against subhuman AI systems and to update them as needed with increasing capacities of AI."

Despite Yampolskiy's concerns and safeguarding ideas, most experts believe that it would be impossible to keep AI a locked genie in the bottle forever. There are just too many behind the scene interest groups, defense contractors and various militaries around the globe spending ridiculous amounts of money willing to rush face first into the rabbit hole of robotics. If that rabbit hole ends up spitting out AI's that have reached levels beyond human scientific understanding and starts to deploy powers such as physic abilities, telepathy or psychokinesis, then Pandora's box will take on a whole new meaning. Yampolskiy warns, "If such software manages to self-improve to levels significantly beyond human-level intelligence, the type of damage it can do is truly beyond our ability to predict or fully comprehend."

Huw Price, Bertrand Russell Professor of Philosophy at Cambridge, Martin Rees, Emeritus Professor of Cosmology & Astrophysics at Cambridge, and Jaan Tallinn, the co-founder of Skype all share Dr. Yampolskiy's views and have co-funded a project together with robotic experts at the University of Cambridge. Where they are currently conducting research into the "extinction-level risks" scenarios that humanity faces with the rise of artificially intelligent robots. With the unstoppable advancements of AI and Skynet plowing full speed ahead, the ability for robots to become self-aware and possibly take over the planet has become a reality. More safeguards are needed, along with openly televised debates and public forums, to help raise awareness on this crucial topic. If not, this reckless drive to mend man with machine at an overboard pace without taking notes of the consequences might ultimately put us on the road to ruin.

Bill Joy, computer scientist and co-founder of Sun Microsystems, wrote a shocking article called "Why the Future Doesn't Need Us" detailing the fall of humanity and the rise of the robots. Now more then 13 years later, his point is more valid and frightening than ever before.

Dr. Roman V. Yampolskiy—*Louisville.edu*

Robot Zombies

6.
Blueprints for Transhumans

The beginning of the 21st century is marked with leaps in specialized technological experimentations of unprecedented proportions. This means that whatever you think technology can do, the reality is unfathomably much more. Our grandchildren and great-grandchildren may not give much thought to how technology changed the world, or how it changed humans—how we have begun baby steps toward a new society that will make the last 200 years seem like the pit of the dark ages. Communication tools, such as cell phones or online games, for example, make it possible for any human on earth—with the exception of those who are not permitted access by overlord governments—to speak, see, interact and even play virtual reality games with other humans anywhere else in the world without the barrier of distance. The downside to this super power is the flood of information available through this ultra rapid exchange. Naturally such an ample field for growth and development has great attraction. What the common man does with the power of information that is accessible to him or her depends on an infinite number of individual circumstances. What the not-so-common man does with such power is extort the common man. Is something happening? You decide—people have never before been as closely observed and scrutinized. There's a lot of spying going on—if it's not to pinpoint your decisions while you're aisle shopping for spaghetti sauce at the supermarket, it's to screen you against terrorists, but someone, somewhere, somehow already knows your name, your blood type and your habits. Is this information necessary? Is it for the greater good of humanity?

Robot Zombies

Which humans are we talking about? Because for all the causes through history introduced for "the greater good of humanity," humans are suffering in greater numbers all over the world.

If we take an honest look around we see *Technology* is not serving us, the common human, well. On the contrary, we are its servants. People follow the path of information, and technology gathers it. It's been over a century since the commercialization of electricity and this 'utility" has enslaved more people than what its harnessing power was designed to serve. The development in robotics has reached the commercialization of unmanned technology; drones for instance, have, in a short period, caused incalculable deaths of innocent people. We humans in our great majority have not mastered our most mundane, earthly elements, and there doesn't seem to be room for fools in this intelligent design because we are dropping like flies. Current humans suffer too much. Our laws restrict us more, and many are outright absurd. In Miami-Dade County Transit (Florida) for example, it is "illegal" to help a person who doesn't have $2.25 fare, by "tapping" the metro entrance stile with their own electronic card, to let in the person in need (mind you, the card is purchased by the individual who uses it; the money in it belongs to the individual), both are criminals and if convicted, become felons for the rest of their lives. In New York City, Mayor Michael Bloomberg banned food donations to homeless shelters, citing that he is introducing food standards. Higher conscience, without the need of external opinion, tells us that helping someone in need is not wrong. Society is shifting, and as it does, it is creating the environment for the new humans.

The present human condition has been years, maybe lifetimes, in the making. With highly advanced technology that doesn't serve us well, and what appears to be an attack on our natural humanist instincts, civilization is moving in a specific direction, one in which there will be little we will be able to accomplish if we continue thinking there is a future for us as we are now. Astrologers say we are traversing energy in space that is propelling us toward what they refer to as, "the end of an era," and which has been hailed by scholars as a "giant leap of consciousness."

Churches claim we are living the end days, the Mayan calendar ended two years ago, and the earth has been moving, shaking and growling. The time will come when the sun will go in a giant explosion into nothingness and take the dust of what remains of the planets with it, thus the extinction of the human species is not a preposterous theory. It may do us well to stop being frightened of the thought of death and think beyond it. It is at this cross of time and technology that humans have turned to various forms of meditations, yoga, sacred sex, hallucinogenic substances, binary frequency audio stimulation, and all sorts of higher conscience initiations in an attempt to understand that we are more than homo sapiens with a DNA sequence with the ability to think and rationalize. We struggle with the question of our origin. Do we subscribe to an evolution theory, or an alien encounter, or both? At some point homo sapiens became humans—an intervention that had early schoolbooks quoting archaeologists and anthropologists citing a "missing link" to account for humans as we are today. Maybe if we all conscientiously got over this useless struggle with our collective identity and opened our eyes to the world around us we would realize one way or another our human ways are dying, and there's no time to waste.

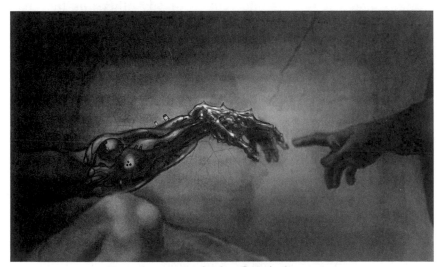

Transhumanist faith—Omnireboot.com

It is possible to be comfortable with our own mortality. In fact, some will say it is necessary or a worthy life would be in vain. The cycle of rebirth happens in front of our eyes, yet we prefer to believe we are exempt from the law of nature. Perhaps this "missing link" connection makes us feel superior to animals, birds, trees and earth in general. It gives one pause to wonder exactly what we gained from the elusive missing link donor? Was it the ability to imagine, to design and create the world around us? Doesn't our ability to constantly create, learn, increase knowledge and reach for the impossible contradict the complacency of accepting our own mortality? Interestingly, the attributes that contribute to our human definition, with some tweaking, can make us inhuman; moving, talking beings without sufficient mental discipline to concentrate long enough to navigate the delicate balance necessary to plan our future. Whether chemically or surgically, over a long or short-term process, the removal of empathy, emotion and the ability, or even desire to procreate, renders the body as a vehicle—a machine, a vessel, like some of our friends call it. Strap on an auditory cyborg hearing device and x-ray vision and it becomes debatable if this body is human. Add bio-human enhancements, and mix in fads like forehead donut implants, piercings and permanent facial designs, and without the intervention of any other link that may potentially go missing, humans are no longer even looking like humans.

The application of computer technology in biology, or biomedicine, is the end of humans, as this civilization has known. The fast growth of applied technology across all fields of science is the future heralded by generations before us. Each new generation is unavoidably forced to live with the results of choices made by the previous generations. It is how the future of our reality seems to be created. We should take time to consider where we're headed individually, and as a species. Researcher, Dr. Aubrey David Nicholas Jasper de Grey, said the person who will live to be 150 years old has already been born. The prototypes of the new human walk among us, and together we are all the bridge to Nietzsche's "overman," or Gilles Deleuze's contextualization

Friedrich Nietzsche at the piano playing Chopin
—*Cambridge University Press*

through "the superfold"—the folding of three parts: a genetic code, information technology, and the interpretation of all three. This superfold Deleuze describes sounds like the more mystical three-part integration of man, like the soul, body and the mastery of both, which results in overcoming the human, and not just becoming a better human, but becoming something else. The attainment of this level of self-awareness is at the core of Transhumanism. Calling it "Spiritual" would smack too much of the unwarranted involvement of religion. Even in philosophy, and by relatively modern masters, our own human desire is to move away from the characteristics associated with humanism—emotion, violence, errors, mistakes; hey, we're only human after all! Deleuze considered himself a Metaphysician, and this is reflected by his affinity for the three parts of man. His early prose work is a challenge in itself, guiding the thinker to new thoughts. It is impossible to cover all areas of technology and philosophy here. This book merely seeks to guide its own readers toward the conclusion of an era, and hopefully shed some light on how to prepare for the next.

This new human, with programmed reasoning power and a longer life span, is not being concocted behind closed doors or in secret lab rooms of power-hungry, lizard-shaped beings obsessed with running the world, nor put together in the bowels of a secluded mansion in a distant, impenetrable island. The development of this new human—the first of many to be sure— had been evolving over a period of time, somewhat slowly, from advancements across all scientific disciplines. Then quickly, as soon as the term "bio-genetic engineering" surfaced in the 1950's among scholars who believed, and rationalized, that the human race is manipulated by technology. The ultimate goal of Transhumanism is to integrate man and technology. That's all well and great, but then what? What will be the culmination of this benefit? Once this integration is achieved and the present human condition begins to change, what will the possibilities of our emerging reality be? To answer this question all we need to do is understand where we stand today with technology in medical science, in military power, in education, entertainment and in philosophy. There will be a day soon when a person who loses a limb will have a mechanical one that will literally attach to message-carrying nerves and enable the person to function exactly as they would normally. Paralysis will be conquered. Imagine driving a car with your mind. Just your thoughts and the engine revs and the car moves carefully out of the driveway. Impossible? How about piloting an airplane with your mind? Jan Scheurmann, a woman paralyzed years ago by a rare disease, can thank neural implants and a DARPA program for the ability to control robotic limbs using her thoughts. DARPA director, Arati Prabhaker, put the implants to the test by having Jan pilot an F-35 simulator using those same implants. The story was introduced to the new America Foundation's first annual "Future of War" conference. This is a major breakthrough for neural signaling, but it also gives us an indication where we could be headed with remote piloting of drones and other aircraft.

Perhaps someday fully self-aware, thinking machines will come standard with our dwellings. What is it we want the future to look like, and what will the beings that emerge after the pruning of humanity be like? Will our own thirst for knowledge be our own doom? The ancient legend of Icarus should be given consideration as an example of the application of mechanical limb technology—a supposition that is not unrealistic considering the volume of data that demonstrates the existence of one or more civilizations with technology more advanced than ours that lived in a time much further back than what most people even believe possible. As we reach new pinnacles in science, researchers find connections with ancient processes that were misunderstood, ignored or forgotten. As the technology for DNA and human cloning manipulation moves forward, the time will come when a mummified pharaoh will indeed be brought back to a "next" life just as his ancient beliefs dictated. In fact, it is probable this may have already happened. What we know today of genetic cloning and nanotechnologies point to a future with some form of eternal life—even if made up more by machine parts than human parts. Nothing visualizes this better than the recently launched remake of

Cyborg from *Teen Titans—Infinite Crisis*

the 80's iconic, futuristic role model of Alex Murphy in *Robocop*. The mass interest in this idea continues to grow and the technology for it is already available—at least to some degree, to common man. The scenes of artificial, mechanical limbs shown in the new *Robocop* are based on real life projects at Chalmers University of Technology in Sweden, where scientists are connecting robotic limbs to the human nervous system of amputees, creating cyborg-like people. This technology will introduce robots in a few short years because the competition to maintain technological edge is fierce. There's nothing that will prevent it. It is predictable that experimentation between biology and technology will create machines with human elements, and humans with machine elements and if it remains, human consciousness will be altered to produce a non-emotional, longer lasting being that will look more human than humans.

The evolution of man is already taking place in front of us. Genetically Modified Organisms (GMOs) are not food. If it were food it would say so. We consume GMOs daily in our coffee, coffee sweetener, bottled water, tap water—these ingredients in everything we have put in our bodies for at least 50 years proved in early animal testing, to produce tumors, among other unnatural growths. Make no mistake that all of us have a substantial amount of chemicals inside interacting with our DNA. How can they not? Essentially, we are mutating right now, as chemicals that reach the core of our molecular structure alter not just the way we look, but also the capacity of our reasoning, and the way we think. We are altered humans already, and as civilization moves toward a future that promises ever-lasting life on a peaceful, green planet, reality is going to be a shocker. Mankind might eventually get there, but it first needs a bridge made up by expendable masses. We are it, and no amount of activism is going to change it. As fantastical and frightening these technological advances seem to some, they are in our near future and we should be considering how new technologies can help us. Individually. The concept of having our brains, our minds, uploaded to a computer that will then create images would be something pretty surreal! Imagine if well preserved because it lay hidden behind another engraved slab that was apparently covering the older, original one. The original engraving depicts what looks like flying machines, among which there is one in the unmistakable shape of a helicopter. The technology we are discovering now existed before on Earth and it is naturally achieved through a progressive cycle much too large for any of it to have survived except in myth. Transhumanism involves the preparation of present humans, intellectually, physically and psychically for what is to come. It will mold the way technology is combined with the human experience. We seem to have been this way before.

We contribute to the future with every action we take today, and thus essentially shaping it on an individual basis and creating a larger canvas as these actions touch others. Clearly, we have a certain degree of influence over the future, but we are not trained, nor is it desirable, that we should envision and plan for our legacy well into a future that will have our descendants playing inter-stellar online games. We have been trained and will continue to be trained to follow instructions and consume.

Trans Cranial Direct Current Stimulation, or TDCS, is a faint electric current that is run through the brain to speed up processes, such as learning a language. This is a technology that fits the goals of Transhumanism. Ironically, this human-enhancing capability, rather than used to foster better communication and understanding among people, was instead used by the U.S. military to train snipers. If Transhumanism is in its infancy and we can change its future, we should be more communicative, and vice versa, with those who are creating these technologies. Who will examine the ethics of enhanced humans, and eternal life? The argument of who gets to live forever and who doesn't—better yet who determines, is one we can be sure will not make it to town hall straw polls. The day it becomes technologically possible to live forever the question of ethics will be the decision of the project's funder. Obviously if everyone suddenly stopped dying the effects would be a global disaster, and if we stopped giving birth that would be an even bigger issue. There's an order of birth, death and

Transhumanist questions—Anne Gordon

rebirth in nature. Even stars in deep space are not exempt from it. Transhumanism, as a school of thought, encourages self-mastery and a strong inner life. Perhaps these principles at the core of each eternal being born from the Transhumanism shift could prevent the future super-human civilization from becoming paralyzed by the super-egos of powerful, wealthy demagogues who won't die. If only the rich have access to technology-enhanced intelligence, every-day Joe doesn't even suspect it, and that takes care of that. On the other hand, just imagine what the world would be like if every person had an IQ of 300. Our entire infrastructure would be different. The world we see today would not be so. We'd probably all be able to build our own transportation at least, treat our own illnesses and have no need for money. It's right about here where the hot and controversial topic of Transhumanism gets its steamy reputation.

There's no question that our near future involves the fusion of humans with technology —and technology is advancing and improving faster every day. Tech news published traditionally is

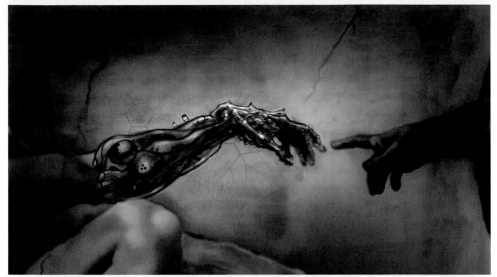
Transhumanist faith—Omnireboot.com

"Rowdy" Roddy Piper in *They Live*—John Carpenter

Google Dreams—Michael Tyka/Google

GOOGLE DREAMS by Michael Tyka

Tokyo Shogun—Michael Tyka/Google

Mad Max—Michael Tyka/Google

Animal Carnival—Michael Tyka/Google

Atlantis—Michael Tyka/Google

GOOGLE DREAMS by Michael Tyka

Trump as President—Michael Tyka/Google

outdated in six months, and every day there's a new announcement in the field.   The Internet is the best example of how technology has changed our lives, starting with immediate connectivity to otherwise unattainable information. Communication technology and advances, like the smartphone and high-speed mobile networks for example, not only have completely changed the way we interact with the world and people around us—it has changed the world. "Flash" gatherings, actual mobilization of people, or instant calls to action is possible today. The instantaneous mobilization of people around the world in real time, or as it is known in mainstream, "live," was used to achieve a change of government in Egypt. That's pretty big. In the history of the world people have never held so much power in their hands as we do now. Imagine if cellphones would have existed in the days of Jacques de Molay! It would be naïve and dangerously unwise to believe those who control technology don't control the world. Yet the great majority of people trust unconditionally those who control the world do so in the interest of the common man. People with power have an entirely different set of values than controlled people. Just as social media was used to create mass opinion during the 2011 Arab Spring, television is used on a daily basis to influence, shape and program what people should think. It is precisely this concept that opened the door to the next obvious step: what are people really thinking? We are a civilized society until disaster strikes, at which point we kill for a place in line to get gasoline. The images from Hurricane Katrina in New Orleans, and the wild panic the city turned into, are still fresh in many minds. Any natural disaster of this scale suddenly brings reality into focus because people are immediately cut off from the false, externally influenced security they have been receiving daily that tells them when to be happy, sad, outraged, or fearful. In a sense, the human behavior obtained from this programming is robotic in nature. For television programs, polls and ratings offer a base from which to stand for competitive financial gains. As technology advances and makes it possible for more information to be processed and calculated, people become unknowing participants in analysis of a grander scale where both visual and audio technology are at play. The frank acceptance of our differences is a realistic first step toward creating the necessary mentality to prepare civilization for the age of cyborgs, when we will see humans with mechanical body parts; robots that will accompany the elderly or the sick, and will be programmed to convey emotion and be able to live virtual lives.

There will be people who will see this as transgression over the creation of man and will not support any movement that intends to tamper scientifically with the human body or the human brain. As radical advances in technology allows us to expand and significantly improve the abilities of our minds, and the forms of our bodies, dramatically altered human beings will become commonplace. Just as the gay and lesbian community pushed, pestered, shouted and paraded their plight of equality into law, so will be the path of the cyborg-human. The hit miniseries, *Orphan Black*, which in its first season captured viewers with its plot of laboratory clones, follows a clear Transhumanistic agenda. It introduces to mainstream the idea of self-directed evolution through human enhancements such as growing a hand, or a tail and of course through cloning. It's difficult to say where society stands on transhuman technology. Average people don't keep up with the fast-paced timeline of scientific developments. When it comes to unknown subject matter, people's opinions tend to side with what they are shown (visually) and what they hear (audibly). It would be irresponsible to believe everything we hear or see in television considering the technology of our times. Western man prefers to be told what to think rather than process, even if slowly, our own internal conclusions. The average person today doesn't have a clue what Transhumanism is, even if the plans for the next human model have been in place for at least the last 75 years. The average individual doesn't get involved because they don't have, or take the time, to consider where technology is going and how to use it best. Billion dollar enterprises however, do get involved and we don't need to search too deep on this one. Google has specific goals in the Transhumanism movement and we will see in the next chapter what the giant of information has in mind. The race has begun. Even if transhuman technology is not immediately embraced at first, in time we will come to accept thoughts or mind downloads as commonplace as we do a cellphone.

Already on websites like Kickstarter.com there are private companies and entrepreneurs looking for funding for projects that will create thought-capturing headsets and haptic feedback suits that will give the individual a more complete virtual reality experience. Virtual sex will undoubtedly be a main point of development, probably followed by virtual sports. Throughout the transhumanist movement there is a strong belief that transhuman technology will end destructive behavior that leads to wars, poverty, famine and diseases. The reality however, points to a starker tendency that could wreak a more negative outcome, especially when the most interested, involved entity in transhuman technology is the military complex, through DARPA. Because of its unlimited access to government funding, the military is able to conduct experiments with developmental technology that never sees the light of day, such as the *Iron Man* suit, once designed with the purpose of protecting soldiers in combat. The philosophical aspect of Transhumanism is one that inspires trust that no matter the mode, the end result would have bettered us all somehow. It is time to think about the future and how we see the world, and ourselves, in this projection. The time has come for the new human and our choices are important. Our minds are being bought and our bodies controlled. As we enter this new age of super-technology, the only way to avoid a wasted life is to unplug from the TV set and tune into the world. Be an example, a shining light to those generations coming behind. Be that first person to live to be 150. In spite of the growing number of unnatural circumstances we appear to be heading into, one interesting thought to consider is this reality, whether virtual or not, includes the dreams and designs of men of purity and integrity, philosophy, faith and science, who believe in immortality. Among these men is one who believes that we will become immortal through the advancements of technology and Transhumanism is a genius inventor, who aims to live forever.

7.
Kurzweil's Quest

Most people have some concept of the finality of death by the time they're ten years old and sometimes earlier. Children will be deeply affected by the loss of a parent or sibling regardless of how much caution there may be in their support group—if any. Being told a loved one is "in a better place" or "they're with God" is frightening to a kid, who up to the moment of being forced to think about death, an event of the future, lived and acted confidently and securely in the present. Who is God? We die? How can it be? What cruel maker would do this to us? Why were we born in the first place? If man contemplates these thoughts persistently over time we develop a certain faith in mankind that becomes the foundation of certain beliefs. There is no cruel maker who has dictated any such fate. There is only energy. Each individual's energy resonates at a particular level that is not exclusive, but overlaps in wave direction. There are many directions and these form a pattern. For example, the plotted vibratory spinning of the earth around the sun when observed over a period of time will show a pattern. Humans are not exempt of this law. Everything forms a pattern. The patterns change with decreased lower vibratory energy or increased higher vibratory energy, and all the ranges in between—the scope of these fields is malleable, moving in waves, and at times some fields are wide and far apart and others are thin and touching. This energy is all around us and we may have been masters of our world at some point in antiquity, but our limited capabilities today find even the thoughts of dimensions, dark matter and quantum realities of existence, incomprehensible. This energy is all around so we can create what we wish. It is a matter of awakening to what we are and what we can do, in this reality at least. Sometimes this awareness happens in steps, struggling from one level to the next, or it can happen suddenly with a life-changing trauma when the accumulation of information in our system and around us is suddenly absorbed and we awake to present reality. The further removed from the original, or source reality, the less likely we are to accomplish 100% duplication of every code so that life for each and everyone is fail-safe perfect for what it was originally intended. There's even a probability all duplications are happening simultaneously in other dimensions!

It's the Ying and Yang wave continuously for everything— but what is everything? According to government physicists headed by Craig Hogan, Director of Fermilab's Center for Particle Astrophysics, we are living in a two dimensional hologram where reality has a limited amount of information, "like a Netflix movie when Comcast is not giving you enough bandwidth. So things are a little blurry and jittery. Nothing ever just stands still, but is always moving a tiny bit." This statement coincides with quantum theory that says everything in the universe is in constant movement so the exact speed and location of subatomic particles is unknown. After four years since testing the holographic theory was proposed, it is finally under way. Whatever the outcome, the goal is to understand the universe and our place in it. Maybe it's a kind of organic, extra-dimensional intuition we access, or something chemically or biologically engineered into our physical lives, but people for the most part have faith in people. The quest for dominion over

Ray Kurzweil shows Stevie Wonder his Kurzweil 250 synthesizer—Getty

our destiny is so intense that scientific technological tests abound, demonstrating this quest. When we're young we have less faith in the unknown and more faith in man. Eternal life is an issue for man—and one man in particular is clearing ground rapidly.

Ray Kurzweil built his first computer program when he was 15 and a synthesizer by the time he was 17, living in Queens, New York in 1965. These inventions earned him first prize at the International Science Fair and made him a national winner of the Westinghouse Talent Search. While still in high school he corresponded with MIT's Marvin Minsky, and although most people's fear of death and desire for immortality are not conversations, or even conscious thoughts in daily living, the man leading the race has been playing with the thoughts for 50 years. Everyone wants eternal life. Medicine, drugs, treatments— whether synthetic or natural are on the rise as new technologies in bio-engineering pave the way in preventive health and healing, curing more fatal diseases than ever before, and regenerating cellular energy. World culture is not tuned for acceptance of death, and perhaps this seeming flaw contributes to criticism of a growing materialistic culture, whose members in great numbers ignore death until it arrives. But just as the great majority of people shelve the thought of death and dying, so do they treat immortality. Even those who consider the subject long enough view it mostly as something realized through works that outlive the mind and body. What we have then is a society subconsciously moving away from death and disease into the realm of immortality—or at least considerable life extension—and psychologically this society is not consciously ready to deal with the speed of the future. Speak with the average person about the merging of technology and humans and they will laugh and call you crazy behind your back. The world is not prepared because for hundreds of thousands of years people lived and people died—some in a short span of years. Their bodies were buried or cremated and the brain and their mind with it. We never envisioned eternal life because there was not enough body of work to show its possibility. This internal dilemma has a physical neurological result, and it may well be related to anxiety in dealing with eternal life.

Yes, there would be chaos if people suddenly stopped dying,

or even if they tapered off slowly with a plague here and a disease there. Just imagine what this would do to the world's economy! Are there enough sources of food and shelter for everyone? What will so many people do? One thing we can learn or gauge by wars is the value placed on human life by governments. It would be naïve to believe that if human life could be extended, it would be available to each individual. When corporate economic forecasters point to economic collapse, companies would put their own family on death row if it meant holding on to a couple of millions. What can we think will happen to us? We live constantly in the future. It is a powerful force of nature that takes place whether we want it to or not and it does not need our consent. We create the future. What kind of physical attributes are we feeding these realities—or is our future destined to be shaped exclusively by those who have more information than we do, in that case leaving people feeling hopelessly out of control? Fortunately, futurologist, author and inventor, Ray Kurzweil, may have this covered and according to him if we just hang in and live long enough, the race to immortality will be conquered by technology and we won't have to wrestle with any of these issues because the future will simply unfold for our benefit. The eventual merger of technology and humans is a theme surrounded with speculations and concerns—which is nothing new for life changing technologies. What is new is the exponential amplification of this technology in the last 20 years and its accelerated expansion.

We are presently living through a major paradigm shift for humanity. Civilization as we knew it is quickly dropping old ways considered relatively modern up until recently, and adapting to the new, which is becoming rapidly outdated. Two hundred or even 100 years ago it was possible for a person to live a life of routine and of a monotonous nature. Agriculture saw the same tools and the same routine season after season. Writing tools remained the same. The delivery of education through a teacher in a classroom environment didn't change. Today technology is developing at such a fast rate that change is constant. When the computer went mainstream with user-friendly language interface software, secretarial and typist jobs, for example, disappeared off the face of the Earth. With programs that correct spelling and grammar errors

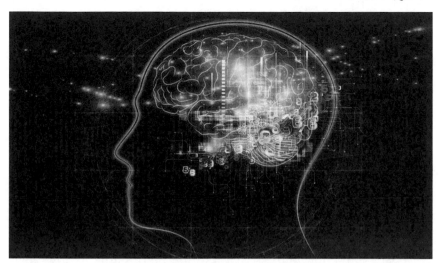

Robot Brain—Sodahead.com

or transfer voice to text, a typist is not necessary and well, a walk to the coffee machine is sensible exercise anyway. Computers have the capacity to make everyone self-sufficient, especially since the introduction of the Internet, and with smartphones we literally have information at our fingertips. This is only how technology has proliferated for the masses in just a few areas. There is no limit to what we can imagine when it comes to artificially creating life through technology.

We don't want to "play" God—again, it is precisely because internally we have utter faith in what appears to be a logical pattern of evolution. Internally, imagination has no limits. It's a kind of living, breathing energy that also surrounds us and vibrates. Some people are seriously in tune with their imagination and the incredible potential of their creativity; others are more concerned with living the day-to-day reality. But it changes and fluctuates and the quality time in our imagination makes the time in reality richer, especially if we use reality to manifest the creative ideas of our imagination. Sometimes we spend more time in one and other times we spend more time in the other—we choose what we want. The universe is energy and it takes a certain kind of vibration to be imaginative. Change is constant; far more than what we imagine, and probably faster, and it all moves to the future. What will we do with this knowledge?

In a previous chapter we discussed the prototype robots that exist as of the writing of this book. Imagine if you will, downloading from a device in your brain, all your memories, personality traits, experiences—all your dreams, nightmares, thoughts, fleeting ideas, absolutely everything you've ever thought, imagined and experienced in this life, storing it in a device one-millionth of a millimeter small—a nanometer—that was introduced through your bloodstream and guided to the hippocampus region of the brain, where it was docked, ready to receive all the data from your mind. This device is then tracked, extracted and kept protected until such a time when technology perfects an Actroid of your choice, or one such human-like machine suitable to act as your new physical temple for your stored memories is ready for you! This technology may develop even faster through the exponential necessities in other fields so that it might be possible to download our memories directly to an Actroid, or freakier still, another human body! Possible?

In March 2012, MIT announced it discovered the location of memories in individual neurons. The reason scientists were able to do this is because they've already located the gene that carries our stored memories. This means the most important part has been identified. After thousands of years of showing up in therapy inventory without a location, science now knows where our memories are! When it comes to artificial intelligence, machines in assembly line fittings may have replaced humans. The heightened level of awareness humans gain from the exponential growth of the rise of the machine has the potential to carry this civilization out of the pressing urgency of the moment to a more compassionate existence. Humans are malleable creatures—we adapt. As we near the point in the future when man and machine will merge, our constant rapid leaps in technology will mean further erosion of our natural energies and adjustments to our human nature in preparedness for when the two shall meet. Whether we like it or not this is happening. Globally, the human psyche is undergoing a reboot. This is the perfect time for someone like Ray Kurzweil to be around with his fascinating computer technology and futuristic visions of immortality.

Kurzweil has achieved things in his life that make for at least 100 lifetimes, beginning with startling accomplishments in his high school years. He designed a software program that synthesized its own songs to different styles based on pattern recognition, a feature that popularized keyboards. How many college students start a company and flip it over for a cool $100,000 in their sophomore year? When Kurzweil was at MIT he started a company using the Select College Consulting Program—a computer program he designed, to match high school students with colleges, which he profitably sold to Harcourt, Brace & World. When tech news around the world announced on December 14, 2012 that Ray Kurzweil was joining Google as Director of Engineering, focusing on "machine learning and language processing," every techie in the world, and probably every investor and economists as well, shifted their perspective.

Those who know Ray Kurzweil personally, or know about him, received the news with butterflies in their stomachs. Kurzweil is a genius, exceptionally familiar with the imagination and the human mind. He has written books on the subject and has even designed computer programs to accelerate the way Wall Street conducts business. He changed the accessibility of music production at an individual level, and created many other technological inventions that have changed our lives—the first flatbed scanner, the first text-to-speech synthesizer, and many other inventions that have earned him honors such as the National Medal of Technology. Kurzweil has found in Google the perfect platform to work on his pet project that he claims to have been tossing around in his head for 50 years: accelerating intelligence that will make us live forever. According to Kurzweil, who is 66, and takes 150 pills a day and a vitamin, enzymes and co-enzymes shot once a week, if you live long enough the day of brain and computer interface is roughly 15-20 years away. Then you can live in a computer forever. He has said that by the year 2029 computers will be able to surpass humans in intelligence. Kurzweil's quest for immortality is in its infancy, but it has stretched a body of work so vast that its impending outcome is a matter of when, and it is no wonder it brought such a huge price tag! The value of this technology will surpass anything we've ever known before. Since Kurzweil joined Google, the Internet giant has been purchasing every artificial intelligence company in the market. Top on the list for an "undisclosed" amount was BigDog's master, Boston Dynamics, the company responsible for creating the relentless terminator-like chase robot, financed by DARPA, for the military. It purchased thermostat maker Nest Labs for 3.2 billion dollars, the British artificial intelligence company, DeepMind, started in 2011, for 480 million dollars, Meka Robotics, Holomni, Redwood Robotics, Bot & Dolly, Schaft, DNN Research and just about every robotics company it can find. And that's not all; Google has also acquired quite an elite lineup of personnel in the field of robotics such as Geoff Hinton, a British computer scientist, and world's leading expert on neural networks, and Regina Dugan, Ex-Director of DARPA, during the agency's renowned creation of the global positioning system (GPS), the stealth fighter and the Internet.

Dugan left DARPA in the midst of an ethics investigation involving conflict of interests related to her own technology company. She is a mechanical engineer with a highly polished businesslike approach. The fact she openly founded her own technology company on the side and met the criteria to be, technically her own contractor, and her own supplier, says more about her than any investigation committee could come up with. Her style impressed Google chairman, Eric Schmidt, and Dugan is now in charge of Google's Advanced Technology and Projects (ATAP) group, whose mission is to deliver life-altering advancements in mobile communications technology. Taking over from Motorola's initiative, Make with Moto, the ATAP group is currently working on Project Ara—an ambitious little tidbit that could have us snapping on and off from our phones, say the contact list, or snapping off the text message module to make room to snap on our infrared night vision. Modular phones that work like fun Lego pieces with a specialized computer program in each. Its functionality meets the basic needs of six billion people, and those of the more arcane individuals. The technology created to accomplish this task follows the path charted by Kurzweil at Google when he spoke about accelerating intelligence across all fields. In medicine, for example, machine-made organs that communicate with other organs, inter-locking mechanical pieces that are integrated with organic tissue, or bone and organ regeneration through 3D printing, are innovations that once seemed fantastic, but today are reachable. Just think of the quality of life these advanced technologies will mean for people who will be living into their 80s, 90s and 100s.

The military is technology's biggest client before any offshoots of a new invention gets to the public. Historically, it is also the military that puts up the stakes, the challenge—the money. Once one or more contractors are onboard, and the project is well on its way, it takes about 20 years, after an invention goes through tests in many applications, before it develops a commercialization edge, if at all. There are thousands of shelved projects and inventions that for one reason or another didn't make it to the outside world, but which are important enough to maintain under patent control. An invisible suit may possibly exist, but really, is it functional outside war tactics? In the last 20 years the public, young and old, has been exposed to some type of advanced technology, and in many cases these technologies are already integrated into daily life, either at home or work. Self-driving automobiles, or a fully voice automated home, are not extreme concepts anymore. When Kurzweil predicts that a computer will behave indistinguishably from a human by 2029, and that by 2045 artificial intelligence will have surpassed human intelligence, becoming the dominant species. Not only is public opinion behind him, based on his track record, and the means Google represents to apply and use the emerging technologies economically and sensibly, it appears he is halfway there. Not knowing what this will mean for humans is a bit unnerving, however, computer scientists of the caliber of Roman V. Yampolskiy shed much needed hope on the subject. He implied during an interview with Nikola Danaylov of Singularity Weblog recorded September 7, 2015 that super intelligence would lack "common sense." What will it mean for us when we reach the "singularity"—the moment that, according to Kurzweil, computer and man will converge and machine will surpass man?

Prior to formally joining Google, Kurzweil had already worked on special projects for Google and he maintained ongoing discussions about technology and artificial intelligence with co-founder Larry Page. Page immediately identified Kurzweil's vision with what Google was trying to do. It had to be obvious to both men that collaboration could mean a place like no other in History. Google offers Kurzweil mammoth scale resources and the fact it's already in the game with the data of one billion people makes Google a neural network. All Kurzweil has to do is to get this enormous global brain to learn. Google is anticipating this will happen and has already put together an ethics board that will examine what it will actually mean for humans when machines are finally learning, gathering knowledge and making their own decisions. It is the exponential criteria, Kurzweil has pointed out, that will pick up a pace sure to make the most tech-savvy aficionado's head spin.

These accelerated technologies will be implemented everywhere from commonplace, every day activities. The way we will be programming our phones and the assortment of interchangeable capabilities it will have, to more sophisticated endeavors such as 3D printing a new home, to absolutely mind-shattering discoveries such as determining if we live inside a 2D hologram. Whatever you think is advanced in technology now, think how much more powerful it can be and chances are we will witness these improvements in a relatively short period. Old ways will die off and some traditions with it. Many people have already experienced first-hand the birth and death of at least half a dozen technologies, such as movie videos and computer disks, that once were necessary to store and deliver information.

Now we can download movies, music and programs directly to our phone, laptop or TV, and explore at our convenience. In communication technology, for example, 3D and holographic imaging are advanced enough to start making a pitch for an enhanced telepresence experience. There are companies such as Holoxica, Ltd. in the U.K. that are already establishing themselves at the forefront of this technology. The case for holograms goes back to 1947 when Hungarian scientist, Dennis Gabor, came up with the concept of holography—a term he coined from combining the Greek words "holos" meaning "whole," and "gramma" meaning "message"—while attempting to improve the resolution of an electron microscope. Because at the time the necessary monochromatic light source from a single point and single wavelength were not available, this development went to the back burner until the invention of the laser by Russian scientists Nikolay Basov and Aleksandr Prokhorov.

Gabor's technique however, was put back in motion in 1962, when University of Michigan's Emmett Leith and Juris Upatnieks duplicated Gabor's method using a laser and holography as a 3D visual medium, resulting in the first laser hologram of 3D objects (a toy train and a bird). In the early 1970s Lloyd Cross developed the integral hologram by combining white-light transmission holography with conventional cinematography to produce moving 3D images. On August 10, 1973 inventor, Robert L. Kurtz, filed a patent that was assigned to NASA and published in June of 1975, for a holographic motion picture camera system (U.S. Patent 3888561 A).

The resurgence of interest in 3D motion pictures and holographic technology means a whole new game in visual imagery. Exponentially, this technology is closer to reality than we think. For example, the Medical School of Edinburgh University uses an anatomical hologram of a six-foot tall human body to show details of the structure. The hologram uses three channels of light sources that enable the viewing of different areas like the internal organs, skeleton and nervous system, and muscle structure as one walks around the hologram. According to Javid Khan, CEO of Holoxica Ltd. at a TED Talk in June 2014, a hologram can be made from anything that is 3D designed with a computer. All this leads back to Kurzweil and his certainty of where computers are headed. At the point in the future when computers have access to all the information in the world, and can recall it instantly, they would have become a billion times more powerful than humans, and will pave the way for mind downloads.

This is what Kurzweil is counting on, and for this purpose he has archived all of his father's writings, because he believes he will be able to retro-engineer him, a-la Spike Jonze's film, *Her*—a story about a writer who creates an avatar of a dead person, based on their writings. A great example of where machines stand today with the gathering of information and the processing of natural language is IBM's *Watson*—a computer with knowledge that was not coded by engineers, but acquired by reading all of Wikipedia. In 2011 *Watson* went on the quiz show, Jeopardy, and not only won, but also exhibited cleverness beyond that of the humans in the game. For example, in the rhyme category *Watson* was given "a long tiresome speech delivered by a frothy pie topping," to which it cleverly beat the humans by responding, "What is a meringue harangue?"

In 2014 Kurzweil explained to *The Guardian*:

> Computers are on the threshold of reading and understanding the semantic content of a language, but not quite at human levels. But since they can read a million times more material than humans they can make up for that with quantity. So IBM's Watson is a pretty weak reader on each page, but it read the 200m pages of Wikipedia. And basically what I'm doing at Google is to try to go beyond what Watson could do. To do it at Google scale. Which is to say to have the computer read tens of billions of pages. Watson doesn't understand the implications of what it's reading. It's doing a sort of pattern matching. It doesn't understand that if John sold his red Volvo to Mary that involves a transaction or possession and ownership being transferred. It doesn't understand that kind of information and so we are going to actually encode that, really try to teach it to understand the meaning of what these documents are saying.[7]

Kurzweil sees technology making us better and smarter in every way. We will outsource certain basic capabilities as we make room for the constant increase in information. One present example Kurzweil has pointed out is how we no longer need to remember phone numbers because we've delegated this to our digital tools. We are moving away from conventional teaching to online learning. We are at the transition point in the education of the next generations. We are learning more, and we are learning faster. One change we can expect in the future is a shift away from traditional public education K-12. Up to now it has been colleges and universities that are responding to the transition toward a teacher-less learning environment by offering all their courses free online. These are all the major universities in the world who are doing this. Granted, if you want to earn credit for a degree, you have to pay for them, but the fact that all this information is freely being given away online is not just a way of ensuring knowledge gets out quickly, but a sign that fresh new mindsets are being sought. We will welcome future nanotechnologies that can travel our blood and set up signals to detect the formation of diseased cells. Cures for disease will be targeted specifically, and within hours the damaged troublesome cells would have received whatever it is they were lacking.

Kurzweil in *Time* magazine — *Time*

There's no need to minimize man a-la Asimov's *Fantastic Voyage* when we can minimize everything else! The medical field in the brave new world is just now getting started, with a plethora of new developments in 3D printing. Regenerating damaged bones or damaged organs, by printing specific areas using harvested human cells, is something all major medical universities are working on, the same way every car manufacturer in the world is already working on a self-driving model. But what of commerce and the new lifestyles we can expect in the future? Imagine 3D printing your own dresses, shoes and handbags! According to Kurzweil, the future humans will not just be smarter, but also immortal. Interestingly, Kurzweil rejects the term "Transhumanism" applied to what he's doing, and says it's because humans will not change all that much in the future. "I've never liked the label Transhumanism, because it implies that we're replacing humanity.

I don't think that's true. What we're doing is augmenting human capability."

Thomas Frey is the Executive Director and Senior Futurist at the DaVinci Institute. He is recognized as a "powerful visionary" and, coincidentally, Google's top-rated futurist. He is the author of *Communicating with the Future*. Listening to Thomas Frey's talks and presentations it may strike us that the basic fundamental method he describes for influencing the future is a simple formula, since we create what we want anyway. The future is our collective creation and we are continuously living in it toward many endpoints. Knowledge is power because those who know, or are "in the know" have a substantial advantage over those who don't know—and let's be honest; the amount of unknown information is much greater than the known information.

Frey speaks in one of his presentations about the evolution of literacy. In our previous era the printing press was a great technology, and literacy was associated with the ability to read, because of the knowledge gained through reading. The printing press made more books available for more people to learn to read and with the ultimate purpose of gaining knowledge. According to Frey it won't be long before people can get a PhD without being literate. This is because literacy has expanded to include computer literacy, and our interaction with computers today is greater than our interaction with books. The knowledge at our fingertips through the use of computers is the new literacy; it is constantly changing and expanding. According to Thomas Frey, the world changed in 2008 with the introduction of the iPhone and the original 552 Apps. From this point on the delivery of information was no longer confined to TV sets.

In an analysis conducted by Frey in 2008, he determined a person received 100,500 words into their heads daily. The information was received primarily through TV (41%), computer (25%) and the rest being minor categories of single digits. As the speed with which we can look up information increases, so does the amount of information humans deal with. Frey calculated every minute there are 700,000 Google searches, 695,000 Facebook entries, 168,000 emails sent, 370,000 Skype calls made, 98,000 new Tweets and 13,000 iPhone Apps downloaded. The consumption and exchange of information is astonishing. The value of information is skyrocketing and according to Thomas Frey the question we should be asking ourselves is how much is our content worth?

In his book, *How to Create a Mind,* Kurzweil touches on the moral challenge for humans to be able to identify with machines that have human-like abilities of language, speech and emotions. Will we identify morally with an artificial person that does not have a biological existence? Good question for all of us to reflect upon. While most of us may associate this issue arising with a domestic robot, in which case the moral concerns may not be any different than what they are now within family law, but how will humans feel toward "spider" robots that act as a search engine in the physical world? While you're wondering whether this is going to happen, don't forget to turn your attention to Google glasses. Overlapping or overlaying a video stream with biometric data within our scope of vision is a technology that is already heavily implemented in media.

It was a matter of time before it was condensed enough to be used in eyeglasses. All the news channels have another 50 stories running below the news anchor. It's not easy on the eyes. It's annoying because these snippets of information compete for our attention as they change, but what if instead these snippets provided useful information for the viewer? [You haven't called Mom since last Sunday. Longest spell to date] [Get more beer right after this play. 13 minutes of ad time.] This is information we will grasp, and perhaps if we're disciplined in our future life of leisure, even follow. Imagine the same technology available with glasses or contact lenses but calibrated to anticipate our judgments and preferences. We will have face recognition, or place recognition from memory, for example. One thing for sure, there is definitely an air of sci-fi sexiness when looking deeply into the metallic, spaceship-like orbs over the pupils of our soul mate's eyes. Nanotechnology will eventually enable a small enough device that can be inserted directly into the retina, perhaps even at birth, and maybe with the added capability of recording a person's entire life. This information will serve the person, hopefully as the primary function of the device, but it will feed the overall system as well—

the global brain of the world. It cannot be any other way. The path of technology we follow already has too much body of work created and invested upon. Meet George Jetson! Grab a cocktail from the instant 3D printing vending machine while we ponder the pitfalls and the advantages of automation technology on the road to merging a more mechanical future super human, and a more spiritual machine.

8.
Cyborgs (Cylons) in Space

With the rapid development of humanoid robots, it is possible, even efficient, that these begin to replace humans at some levels, especially in situations that put human lives at stake. However, the fear that humanoid robots will appear in more complex business professions, like stockbrokers and architects, is not completely unfounded. According to Tokyo robotics pioneer Hiroshi Ishiguro, human looking robots will replace pop stars within a decade saying:

> There are many possibilities, for example an android robot is more beautiful, probably, and they never get tired and can keep singing songs forever. And how about a fashion model? Or newscasters and receptionists—even famous movie stars can be computer generated. Androids never get old and so you can keep a young identity by creating an android that will last forever.[8]

No longer would James Dean have to die young to live forever when an android could do the same thing without having to die in a terrible car accident. Ishiguro gained worldwide acclaim in 2010 after creating an android version of his own self that carried out his duties as a lecturer and professor at Osaka University. His lab is on the forefront of creating lifelike humanoid robots designed to be indistinguishable from their human counterparts. "People watched my android more than they would watch me, I think it's a good way to keep my authority and my identity. Androids never get old, so I'm ready to die."

Hiroshi Ishiguro and his Android—*Robotronica*

Ishiguro's humanoid robots are already mind-blowing, lifelike enough in appearance to be mistaken for a human, but the robots brain-interface is still a long way from completing the human process, despite the advancements made at Google's brain development programs, and recent reports that scientists have created brains via stem cells in petri dishes. The new TrueNorth chip from IBM was modeled after the human brain and is the fastest computer-based software to emulate the brain's activity, but still operates at least 1,500 times slower than our brain does. Brad Borque of *Digital Trends* writes:

A computer needs so much power for this function because it is astounding what the human mind is capable of, even passively. Actions like seeing a ball coming toward you and catching it are easy for you, but programming a computer to do all of those things is a complex circus of calculations and responses. The goal of the project is to create a computer that can simulate vision, and take action in response to stimuli and memory, and TrueNorth is the closest technology has come yet. The TrueNorth chip is part of an effort that began at IBM in 2008, with the goal of producing a new form of computing architecture based on the brain's neuron and synapse network. Rather than simply running calculations across 4 or 8 cores as quickly as possible, the TrueNorth chip recognizes patterns and leverages them for more efficient data handling and processing, with the help of its 4,096 cores. These cores aren't like the cores found in your home processor though, and they're used quite differently. Rather than running all the time, each core is called to operate when needed, keeping heat and energy use down. All of the cores communicate across an event-driven network, which also means that the system can scale, cores

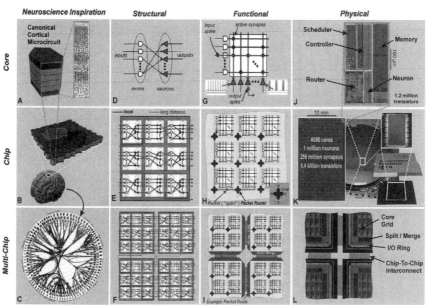

Diagram explaining the various aspects of IBM's TrueNorth chip
—*Extremetech.com*

can stop functioning, and everything will continue to operate without interruption. Those 4,096 cores working in harmony represent one million neurons and 256 million synapses. That's still a far cry from a human brain's ten billion neurons and 100 trillion synapses, which is the ultimate goal of the IBM cognitive computing project. These computers, which would take up less than two liters of space and consume only one kilowatt of power, could revolutionize everything from transportation to vision assistance. It's a totally new direction for computing, unlike any related innovation we've seen in 70 years.[9]

These advances in computing brain developments go hand in hand with Hiroshi Ishiguro's current six-year project aimed at creating emotional states in the robots through intention and desire. Ishiguro believes the project underway at Osaka University is the gateway to creating an empathetic robot. Ishiguro says, "For example deep learning. If we use deep learning we can have very much human-like visual functions, auditory functions and probably in the near future the robot can have a human-level intelligence. Right now the biggest challenge is to implement intention and desire. Then the robot can have a more complicated internal state and mind."

With robots becoming more lifelike, one has to wonder if the scenario's played out in the fictional sci-fi television series,

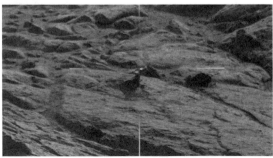

Alien Robot Arm Discovered by Mars Rover—*Gizmodo UK*

Ancient giant robot?—*Flipboard.com*

Battlestar Galactica, will become a reality in the distant future? Or maybe it has already happened before in our galaxy, as the red planet Mars looks to be littered with ancient technology left over from an advanced unknown civilization? Some of the objects that appear in the images being sent back include ancient fossil bones,

THIGH BONE ON MARS?

OBJECT COMPARED TO DINOSAUR THIGH

Thigh bone on Mars?—*Exopolitics.org*

Martian ruins with a pyramid in the distance—*NASA JPL*

pyramids and some even look to be robotic in nature, including what appears to be an android's disconnected arm and wayward head, left to rest on the dried primeval Martian ground forever.

As the new space race gears up for an eventual manned flight to Mars, the question must be asked whether robots or humans should be making the journey? You can send a robot to Mars for the same cost as making a movie about sending Bruce Willis to Mars. The odds are that any upcoming, publicly-staged manned mission to Mars will likely be directed by Spielberg on a secret soundstage than happening for real out in space while androids are used for the real deal.

This scenario was foretold in the forgotten 70s movie, *Capricorn One,* starring O.J. Simpson, about a faked government mission to Mars filmed on a soundstage in Nevada. The complications of sending humans into space is well known and it seems highly unlikely that humans could withstand the six-month voyage through deadly waves of outer space radiation on their journey to Mars. Sending a fleet of robots to Mars sounds like an efficient and logical choice before real people are transported for permanent relocation.

Perhaps the biggest robot presence in space is on the ISS (International Space Station) home of the telepresence robot, Robonaut 2, which made a recent appearance in the One Direction hit music video *"Drag Me Down."* ISS is also home to advanced stem cell experiments, which begs the question: whose genes are up there seeking human immortality aboard the ISS? This puzzle is closer to becoming an episode of *Battlestar Galactica* than an event of scientific proportions. In *Battlestar Galactica* Earth was destroyed and made an uninhabitable wasteland by a series of nuclear wars waged by its out-of-hand robot creations. These robots were called Cylons, and some were so lifelike and advanced it was impossible to tell them apart from humans. In fact, some of the Cylons weren't even aware that they were such. The original Cylons were reptilian; who soon realized their race was becoming extinct and diligently engineered a new race of robotic Cylons to replace them, programmed with the original's software—something akin to consciousness in human terms.

The origins of the reptilian Cylons are unknown, other than they are a warring, conquering race lusting after control of

Cylon from the original *Battlestar Galactica* series—*Nothinguncut.com*

113

vast galaxies. The choice of the reptilian theme, one of the most controversial conspiracy theories in the world, shows up as the source of the Cylons origin in *Battlestar Galactica*. These original reptilian Cylon's maintained control by creating sophisticated robots and then eventually self-aware intelligent cyborgs. These machines were built with titanium armored strength and potent computational capacity. The Cylons believed that humans upset the balance of the galaxy and considered it their mission to destroy all forms of intelligent human life in the universe. Armies of Cylons swept through each sector of the galaxy to the point of becoming machines far superior to their creators, and ultimately destroying themselves in the process. The remaining Cylons eventually evolved to emulate the human form. In the 1978 premiere episode, Captain Apollo explained to his son the origins of the Cylon's mission and why it was important that humans survive:

> They're not like us. They're machines created by living creatures a long, long time ago...a race of reptiles called Cylons. After a while the Cylons discovered humans were the most practical form of creature in this system. So they copied our bodies, but they built them bigger and stronger than we are. And they can exchange parts so they can live forever...There are no more real Cylons. They died off thousands of yahrens (years) ago, leaving behind a race of super-machines, but we still call them Cylons.

Battlestar Galactica paints a dark picture of future relations between humans and sentient robots, one that sees the Cylons overthrowing their creators. As we make progress in the fields of robotics and artificial intelligence ourselves, the fear of being stomped out by our space-traveling Cylon creations could become a real concern in the not-too-distant future. Supposing humans send a fleet of superintelligent robots to Mars and their process of learning superseded human intelligence by a thousand

Sony VR—*Kitguru.net*

fold? Creating themselves into any outer form they wish? Would they see humans as their creators, or as a race to be subjugated? The seeds to Earth's moral and physical destruction are already being sown, much like they were shown in the series *Caprica,* the prequel television series to *Battlestar Galactica.* The citizens of *Caprica* were much like we are today, dazzled with enlightened technology and on the forefront of a robot-inhabited dystopia. The theme of virtual reality was a prominent one in the series as a glasses-like device called the "holoband" was used to transport people into virtual reality clubs that catered to every possible situation and deviation imaginable to a software programmer. Simulated hardcore pornography and human sacrifice, clearly indicating *Caprica* was a world of withering values and vanishing morals—sadly, a world much like our own. We've even begun to harness the power of virtual reality to the point of Sony announcing a new video game console inspired by *Battlestar Galactica's* "holoband!" Some Wall Street experts are counting that virtual reality is the next frontier for retail.

But there's more to virtual reality than just wandering around a simulated 80's Miami disco scene and spending bitcoins on virtual blow. There's the concept of whole brain emulation, meaning a way for someone to copy his or her own mind and then uploading it into the simulation, thus creating a virtual clone. Whole brain emulation involves the complete scanning and mapping of a biological brain in detail and copying its state into a computer. In virtual space people will be able to live within the simulation, where the mind would reside forever within the virtual simulation's matrix-like architecture. This notion is not new. Philosopher Nick Bostrom has already proposed that our reality is a simulated hologram, coded by higher intelligence and living in a simulated universe.

The possibility of mind transference is profound and calls into question issues of the permanent and absolute self, along with broader philosophical and scientific predicaments about how such a technological feat could ever be accomplished and what it means to the human soul. These questions, and the implication of physics and philosophy ultimately lead to the wonders of nanotechnology and Transhumanist, robot zombies.

9.
Nanotech Dreams

Nanotechnology is the engineering of functional systems at a molecular or atomic scale. Humans can't touch and separate atoms because we are too big, but creating technology that makes it possible to construct at an atomic level is what nanotechnology, in its original sense, is all about—building complex things from the bottom up, with atomic precision. The word "nanotechnology" is attributed to K. Eric Drexler when he described constructing machines the size of molecules. The term caught on and has since been adopted by the U.S. National Nanotechnology Initiative, to mean anything with original properties that are smaller than 100 nanometers. Twenty years earlier, Richard Feynman, the brilliant Nobel Prize winning bad boy of theoretical physics, explained in an interview recorded during a family trip in Great Britain the exact same concept:

> I want to build a billion tiny factories, models of each other, which are manufacturing simultaneously... The principles of physics, as far as I can see, do not speak against the possibility of maneuvering things atom by atom. It is not an attempt to violate any laws; it is something, in principle, that can be done; but in practice, it has not been done because we are too big.[10]

We may not be able to move tiny atoms with our index finger, but they can be moved and changed through chemistry, engineering chemical reactions in a biological system—and this brings us to the mind-boggling mathematical task of creating precision-guided mechanisms that can chemically alter or stimulate cellular molecules. Already working on computing software and algorithms that will help achieve this are Battelle, the world's largest nonprofit research and development organization that also manages the world's leading national laboratories, including U.S. National Laboratories, and the Foresight Nanotech Institute. Battelle's mission is to "solve what matters most," and one look at their website's home page will get right to the point with a short, groundbreaking video that shows the application of neural bypass technology in the form of an implanted chip that helps a man move his paralyzed hand with his mind. The procedure, described by research leader, Chad Boute, is similar to a heart bypass, but instead of bypassing blood, bypasses signals. Focusing on general purposes, nanotechnology will affect manufacturing and will change society in a big way by creating better, more efficient products across all industries. The U.S. *National Science Foundation* put it this way:

Transhumanism and Nanotechnology—*Youtube*

118

"Imagine a medical device that travels through the human body to seek out and destroy small clusters of cancerous cells before they can spread. Or a box no larger than a sugar cube that contains the entire contents of the Library of Congress. Or materials much lighter than steel that possess ten times as much strength."

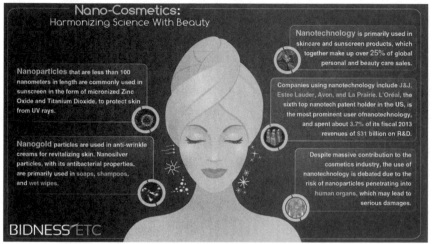

Nano-cosmetics—*Bidnessetc.com*

Soon at our fingertips will be a "nanofactory"—an all-knowing robotic device that will compute, process and administer chemicals that will produce a pen—for example, or any other item. With exact precision this machine will be able to produce items, inexpensively, directly from the input instructions. This otherwise dreamy gadget obviously has the potential for abuse and misuse, and when taking into account the speed at which these nanofactories will replicate themselves, creating sensible policies, is not something to scoff at. There's a hot debate about how soon the exponential growth of technology will catch up to the point where machines will be building machines in record speed. There are computer scientists who say this will be a possibility in 20 to 30 years from now. According to top tech news source, CRN, however, this may happen within the next decade thanks to the leaps in enabling technologies, such as optics, nanolithography,

Richard Feynman—*Elcaminantefractal.wordpress.com*

mechanic-chemistry and 3D prototyping. Whether related or not, the point in time when advanced technology broke into the mainstream, and has been exploding exponentially since in every area of living, points to the late 1940s after the alleged alien crash in Roswell, NM. It may be connected or not, the point being the timeline. In the last 70 years the world has changed dramatically fast, with advancements in nanotechnology leading the way. Take for instance the nanoparticle-filled ink spots that can conduct electricity.

Researchers at the University of Illinois at Urbana-Champaign (and many other teams) are making conductive ink from silver nanoparticles, which are then shrunk using acid. The nanoparticles are suspended in a cellulose solution so they have a greater viscosity and can flow from a pen, quite literally. A line drawing becomes a silver wire that can carry a current, enough to power an antenna or even a small LED display, like the light bulb at the top of the house in this lovely drawing. The pen allows circuits to be embedded on uneven surfaces, and it enables a new type of creative design. "Everything, when miniaturized to the sub-100-nanometer scale, has new properties, regardless of what it is," says Chad Mirkin, Professor of Chemistry (and Materials Science, Engineering, Medicine, Biomedical, Chemical

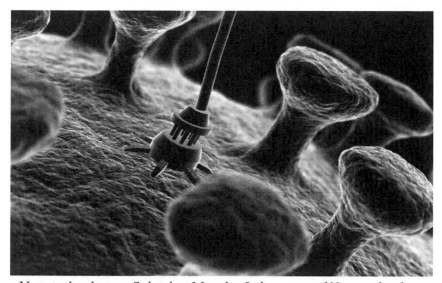

Nanotechnology—*Salvador-Morales Laboratory of Nanotechnology*

and Biological Engineering) at Northwestern University. This is what makes nanoparticles the materials of the future. They have strange chemical and physical properties compared to their larger-particle kin. The thing that matters about nanoparticles is their scale. Nanoscale materials are used in everything from sunscreen to chemical catalysts to antibacterial agents, from the mundane to the lifesaving.

"I spilled wine at a Christmas party once, and I was terrified. Red wine on a white carpet! And it wipes right up," Mirkin recalled. "The reason is the nano-particulate used to coat the carpet keeps that material from absorbing into the carpet and staining the carpet."[11]

The reach in nanotechnology in various different fields are vast and range from nanoscale assays used to screen for cancer, infections and tumors, to gold nanoparticles juiced up with DNA to detect certain types of bacteria and diseases in a person's bloodstream. Mirkin explains:

> The smaller you go, the ratio of surface to bulk atoms goes up. At a larger scale, the atoms at the surface are relatively inconsequential. But at nanoscales, you could

have a particle that is almost all surface. Those atoms begin to contribute very significantly to the overall properties of the material. Much of it is going to be embedded in conventional products that we buy and don't even think about. There's nothing inherently good or bad in terms of making things small…you arrange a simple molecule in a spherical form, and it naturally enters cells better than anything known to man. That is a paradigm shifter for how we think about creating new therapeutics—in this case, involving the world's most important molecule, and learning how to arrange it in new forms on the nanoscale. The issue ultimately is what do they do, and what are they used for? Given the application, have we considered the proper safety analyses and implications? And so far, I think we've done a pretty good job.

The groundbreaking research going on in the field of nanotechnology is happening at an alarming rate, and the general public, in its greater majority, hasn't a clue. Practically every week, scientists announce a new ability of nanoparticles to deliver genes, drugs or chemical messengers inside cells. Nanoparticles of different shapes and chemical makeup can track and target a specific cell of a chemist's choosing, and perform a variety of tasks. This nanotech revolution can change our world in the blink of an eye. Imagine a world where you can easily access energy from the quantum scale. Factories will become ghost towns as nanobots create whatever form of energy is needed from the periodic table to instantly purify our air and water. Gone are the days of toxic landfills and endless consumerism as a newfound nano-utopia fosters an environment where most diseases and genetic problems will not exist. But it seems that the true benefits of nanotechnology are being kept hidden to the selected elites who are hoarding it for their sinister transhuman purpose of gaining eternal life. The non-elites are being directed down the less than glorious Tanshumanist path that aims to merge man with machine and create self-aware robots with super brains. Our culture is littered

Nanotechnology—*TLV1*

already with synthetic food, synthetic voices from untalented "singers," synthetic body parts, synthetic memory implants and synthetic everything else. We are shredding our humanity by the minute with the hopes of replacing it with the latest technology. This "becoming" society that merges man with machine is straight out of the book of *Revelations* 9:6 states:

> *And in those days shall men seek death, and shall not find it and shall desire to die, and death shall flee from them.*

The dawn of biological computers is upon us as Stanford engineers have already created biological transistors providing the missing components necessary for the creation of a biological computer; programs function of storage, transmitting information and logic on a cellular level, replacing microchips with advanced DNA-infused proteins. The cyborg courtship continues with the advancements made in a 3D printing, where bioprinting is on the verge of being a billion-dollar business within a decade, and the hopes of the limbless high, as soon they will be recipients of freshly printed cybernetic arms for half of today's costs. Cosmetics giant L'Oreal has partnered with biotech firm, Organovo, with the hopes of developing 3D printed skin tissues. The French cosmetics company has already been growing skin in laboratories for decades, as it is illegal to test cosmetics on animals in Europe. Procter & Gamble has also been inspired recently to invest in 3D printing, as the consumer goods giant announced it was holding an academic competition and issuing large grants to qualified researchers. While the push for the Transhumanist merging is being rolled out, the obvious first signs of its arrival are zombies. Not exactly *Walking Dead,* eat your brains out zombies, not yet anyways, but technological zombies staring blankly at their iPhones unaware that they are being decoded and reprogrammed.

Nanotechnology — *Gizmodo*

Robot Zombies

10.
Zombies and the Living Dead

Zombie lore can be traced to the 10th Century BCE account *The Epic of Gilgamesh*, considered one of the oldest written works in human history, and which includes a battle with armies of flesh-eating undead. Today we can thank science fiction, Voodoo, and TV for the conceptualization we have about Zombies — people who return to life after being dead for a brief period of time exhibiting bodily animation and an appetite for flesh. This insatiable appetite apparently produces such a high level of pleasure that they will do anything for it, including returning from the underworld. Obviously nobody comes back from the dead, at least that we know of, and especially not to eat flesh, but the concept of mindless behavior and dead flesh is not entirely false. First though, let's review what the sings of being a zombie are. The main one is of course, being undead, and this we know has no real footing in medical terms. There are however, diseases, and effects of certain neurological toxins, that can make people act like a zombie. These include symptoms such as rotting flesh, a state of mental trance without cognitive function, lack of ability to speak or communicate, a slow walking pace, and a taste for human flesh. Sleeping sickness, rabies, necrosis, dysarthria and leprosy are some of the diseases recognized and categorized by medical science.

Sleeping sickness is caused by a parasite called Trypanosoma brucei and it is transmitted by the tsetse fly, prevalent in Africa. BBC interviewed Professor Sanjeev Krishna of the University of London when he was a doctor in Angola, and explained the onset and symptoms of the disease were headaches, aching muscles and perhaps itching. In the late stages of the disease when the parasites have reached the brain, the victims find it hard to concentrate and become irritable, with slurred speech and they stop eating. They don't sleep at night and can't stay awake in the daytime. Eventually they are unable to perform any simple task and fall into a trance-like catatonic state, and die. There are no vaccines or any other means to prevent infection once the tsetse fly bites a person. Sleeping sickness kills over 70,000 people every year. Researchers at the Sanger Institute have sequenced the genome of the parasite that causes this disease, and while this is hopeful for its treatment, the knowledge of how to manipulate it carries the danger of biological weaponry. Sleeping sickness however doesn't make anybody want to eat people, but rabies, a virus transmitted by infected animals, causes massive swelling of the brain, and comes pretty close to exhibiting the zombie-like desire for eating flesh. Rabies is most prevalent in Asia and Africa. Although Louis Pasteur successfully treated a rabies infection of a child through a process that opened the door to our modern vaccinations, an estimated 55,000 people die annually from rabies because the vaccine needs to be administered before the onset of symptoms. These include uncharacteristic strange behavior, inability to communicate or walk, aggressive attacks on people and the loss of reason. Human to human transmission of the disease appears to be rare, if at all possible, since we don't see people going around biting other people.

Necrosis means death in Greek. In the case of the disease, it means the death of cells or groups of cells. Necrosis is considered a condition, rather than a disease, without an actual known cause. A person afflicted with necrosis is the closest we can come to the physical manifestation of what a zombie would look like since the patient's body tissues are literally dead, even though vital organs are functioning, and therefore alive. The spread of the infection, or whatever it is that causes necrosis to healthy cells can spread over a large enough area to be fatal on its own. Necrosis isn't contagious and unlikely to trigger a spontaneous zombie-like outbreak.

Dysarthria is a disorder that affects the control of human speech and it is neurological in origin, which means it touches upon the brain aspect of the zombie legend. The disease is characterized by a malady of the nervous system that impairs the control of the tongue, lips, throat and even the lungs. This causes the patient to experience a struggle to articulate, producing sounds like grunts and moans of uneven pitch. While this condition on its own doesn't warrant zombie comparison, coupled with, let's say, necrosis, rabies or sleeping sickness…you get where we're going with this.

Again considering decaying, rotting flesh as a common trait of zombies, we can see why a disease like leprosy, which goes far back into antiquity and is still around today, can be a culprit in the inspiration for zombie stories. To be clear, leprosy doesn't cause body parts to rot away and fall off, but the spread of skin lesions over parts of the body give a decaying appearance. In addition, the damage caused to nerves manifests in numbness and lack of sensation that if leprosy were to affect limbs such as the legs; it would cause the slow, labored gait we associate with zombies. Leprosy, once one of the most stigmatized and feared diseases in history, the same terror we associate today with a population of dead people rising to eat us, is under control with almost 100% of people naturally immune to it and over 15 million cured of the disease in the last 20 years. In the 2007 book, *The Colony: The Harrowing True Story of the Exiles of Molokai*, author John Tayman tells of the lepers in this Hawaiian colony and how they were literally considered and treated as if dead, even if they showed every other sign of life, including reason. The spouses of these victims of leprosy were granted immediate divorces on the basis they were essentially married to dead people. The sufferers were exiled to a remote island and were left to die.

Aside from these diseases that have zombie-like symptoms, there are other ways to produce populations of non-reasoning, non-thinking automatons that don't have one single synthetic wire running through them. Some people may argue that today's American society is brain-dead and frankly, a look at popular clips produced by late night TV hosts or talk show personalities, where an interviewer walks among the public and asks them reasonable questions that any normal reasoning person should have the answer to, can make one wonder if we are not in fact living among zombies? One of the most loved and respected journalists of American television, whose rise to prominence came from his wartime reporting of WWII, Edward Murrow, was personally involved in the struggle between networks and the control corporate advertisers began to exert over public information. Below is his famous speech, delivered to the Radio and Television News Directors Association in Chicago during the fall of 1958:

> This might just do nobody any good. At the end of this discourse a few people may accuse this reporter of fouling his comfortable nest, and your organization may be accused of having given hospitality to heretical and even dangerous ideas. But the elaborate structure of networks, advertising agencies, and sponsors will not be shaken or altered. It is my desire, if not my duty, to try to talk to you journeymen with some candor about what is happening to radio and television. And if what I say is responsible, I, alone, am responsible for the saying of it.
>
> Our history will be what we make it. And if there are any historians about 50 or 100 years from now—and there should be preserved the kinescopes of one week of all three networks— they will there find, recorded in black and white and in color, evidence of decadence, escapism, and insulation from the realities of the world in which we live. We are currently wealthy, fat, comfortable, and complacent. We have a built-in allergy to unpleasant or disturbing information.
>
> Our mass media reflects this. But unless

we get up off our fat surpluses and recognize that television in the main is being used to distract, delude, amuse, and insulate us, then television, and those who work at it, may see a totally different picture too late. I began by saying that our history will be what we make it. If we go on as we are, then history will take its revenge and retribution will not limp in catching up with us. Just once in a while let us exalt the importance of ideas and information. Let us dream to the extent of saying that on a given Sunday night, a time normally occupied by Ed Sullivan is given over to a clinical survey on the state of American education. And a week or two later, a time normally used by Steve Allen is devoted to a thoroughgoing study of American policy in the Middle East. Would the corporate image of their respective sponsors be damaged? Would the shareholders rise up in their wrath and complain? Would anything happen, other than a few million people would have received a little illumination on subjects that may well determine the future of this country—and therefore the future of the corporations? To those who say people wouldn't look, they wouldn't be interested, they're too complacent, indifferent and insulated, I can only reply: There is, in one reporter's opinion, considerable evidence against that contention. But even if they are right, what have they got to lose? Because if they are right, and this instrument is good for nothing but to entertain, amuse and insulate, then the tube is flickering now and we will soon see that the whole struggle is lost. This instrument can teach. It can illuminate and, yes, it can even inspire. But it can do so only to the extent that humans are determined to use it towards those ends. Otherwise, it is merely wires and lights—in a box. Good night and good luck.[12]

Edward Murrow warned of the power of mass hypnosis through television. Since then generations have been raised by this medium, becoming the nanny of many young people while the parents became complacent. Today we don't have just three networks. Today one can spend hours just clicking through a menu of thousands of selections and the same that was true in the early days of television is now exponentially more so. We are being brainwashed. Brains under the control and manipulation of others are more of a zombie trait than having them for dessert. How, you might ask, with a hint of indignation, since you are perhaps sitting on your couch, without a dripping faucet slowly creeping into your subconscious to irritate you. Of all the available methods of deprogramming and reprogramming, the mass media is able to manipulate its audience using techniques designed to bypass the

Edward R. Murrow — *Life*

"thinking" part of the brain, reaching deep inside to the button that says, "Obey."

The visuals are important, so yes, sex and hamburgers will be glorified, but there are many more ways to get you to comply, such as getting you to chant slogans. Sound familiar? Teachers, church leaders, life coach presenters and politicians are taught that the way to silence the single, doubting voices in a crowd is to get them going on what is known as "thought-stopping techniques" by getting them to chant a repetitive slogan. Why? Because it works. The analytical part of the brain and the repetitive part don't live in the same block. It is not possible to solve a complex problem requiring logic, while forced to scream a slogan or a chant at the top of our lungs. Some chant examples: "President X! President X!" or, "Say it with me people!" Even techniques that are meant to help us adjust or heal use the same method, such as meditation and prayer. They work because they shut down the thoughts, or the internal voice(s) that many people have, and don't necessarily mean they shouldn't. Whether at football rallies, Wrestlemania or political meetings, be alert and notice how audiences are trained to fill any silent gap with loud chants. Beware of slogans. They are used to brainwash you into believing and obeying without thought.

What about our subconscious? How can messages, especially nonsensical or illogical absurdities embedded in our subconscious mind? Headlines, headlines, headlines! All the major news agencies thrive on it. Our opinion is manipulated through headlines since most don't have the interest or the time to even read the stories, we have a quick look through the headlines. Based on the way our brain stores memories, this technique is only concerned with planting an idea. That's all. If we see a headline that reads "Ron Paul Made Racist Remarks in Newsletter" this is the idea that is planted. Even if the headline is explained in the actual article revealing it was not Ron Paul who made the remarks, but the editor of the newsletter, it won't matter because the idea was already planted and the next time we hear the name "Ron Paul" we will automatically attach it to the headline. We will say, "oh yeah, that's the guy that made racist remarks." We would be wrong, and we would have been used to mislead others. This is called source amnesia, and the technique relies on the fact we will know we read this somewhere, but we won't remember where. This is an easily exploited mechanism because our brains have storage limits. The overload of information coming at us today contributes to the problem itself. Even if the headline is not true, the repetitiveness of it causes us to only grasp the part intended for us to absorb. The only thing we can hope for is to be aware of these traps, and question why it is that all hundreds of news channels catering to a populace of half a billion, say exactly the same thing. Try it.

Another way of manipulating our beliefs and actions to incapacitate, or significantly reduce our thinking, is by restricting what we watch or read. A good example of this can be seen in religions that restrict or condemn information not specifically related or supportive of the religion's belief or custom. Religions as distinct as Islam and Evangelical Christianity, for example, both restrict their members in the material they read. They place emphasis exclusively on reading either the Quran or the Bible, and personal accounts specific to each religion, undoubtedly to preserve control over their followers. Scientology goes as far as filtering the Internet access of their members. By preventing contact with the opinions of others outside the circle, any influence that may awaken self-thought is effectively curtailed by creating the illusion that beliefs outside those of the specific religion, cult or church are harmful or evil. Most people want to be good. They follow these rules without question. This technique is not just limited to religions and cults. Studies have shown that our human brain is wired to receive a jolt of pleasure when we read or see things that support or agree with our own beliefs, and from discarding information that disagrees with us, even if it is thoroughly explained or documented. This type of brainwashing response is classically evident during presidential campaigns where people will side with a candidate without the slightest reasonable notion other than: "She's a tough cookie." The overwhelming amount of news outlets, each lambasting the others, creating watchdog groups for the purpose of just getting dirt on each other, teaching us to not only ignore points of importance, but to talk right over them as if they were insignificant or stupid. This is an effective technique that keeps people from coming together through logic and common sense and allows triumph through division.

Are you ashamed of your opinion? Are you hesitant in speaking your mind? It's called keeping you under control through humiliation. A valuable technique that appeals to the fallacy of ridicule is based on using something as simple as repeating something with a tone of disdain. If an idea can be portrayed as ridiculous, it is unlikely anyone will invest any time in analyzing it to self-determine whether or not it is. This method can stop anything in its tracks because the logic behind it is that if one considers something that's ridiculous it would make that person ridiculous, and by golly, we are not ridiculous! Believing in ghosts, whether you have seen anomalies or not, is simply ridiculous. There's no better way to shut people up if they're shamed. It works because our brains have something called amygdala that controls emotional reactions, and when it is stimulated, can shut down the analytical side of our brain. Mockery is an effective mode of enforcing silence and conformity. Just watch fourth graders in the schoolyard.

Given an either/or choice, no matter how illogical they are, in fact the worse they are, the better, because now you are convinced to pick the lesser of two evils. Says who? "You are either with us, or against us." There is no choice in a statement like this. Coupled with slogans and limited access to outside opinions, the black or white choice is an effective manipulative technique that is used most prominently by politician's intent on pushing agendas that are not in line with public interest and must be disguised in order to garner needed public support. This method appeals to one of the most basic foundations of the human mechanism, "fight or flight," which causes us to make quick decisions by bringing down every situation to these two choices. To bypass our critical thinking circuit, there is no easier way to do so than to make you scared or anxious and limiting your time to make a decision. The brain's analytical neocortex is where we scan a situation to determine the right or the wrong action. The more stressed we become, the more willing we are to attach to those who feel like we do. This is an ancient mechanism of survival, wired into our brains by our tribe-forming evolution that at one point was necessary for survival against a hostile environment that we did not have the full capacity to dominate. This technique is what allows us to criticize, ridicule and silence any belief outside of ours. Because there are others who feel the same way, we will believe that our convictions are the correct, and the only ones, that will save us from destruction. The result is the widespread division of opinions in large enough quantity to support any action deemed necessary by the controlling entities, be it corporations, banks, energy, pharmaceutical companies or alien super intelligence. It's either us, or them—the ultimate sign of control. Nothing can exemplify this attitude better than "The Samson Option"—an option that Israel has made clear it will use in the event of an imminent threat of a substantially crippling attack, by unleashing nuclear missiles aimed at every major city in the world, including the most populated cities and the capital of the U.S., their strongest ally.

With all these scenarios aside, humans are also being tampered with physiologically and neurologically through toxins in the water we drink, the genetically modified organisms that form food-like products, the poisons used as pesticides in the food that is grown, and the chemicals introduced by means of radiation and chemotherapy-like treatments. A study conducted by Harvard University concluded that fluoride is a neurotoxin and affects our brain functions. Fluoride was given in heavy doses to prisoners in German concentration camps. It pacified the mind into acceptance, again curtailing critical thinking and was the main reason why hundreds of people didn't rebel against a handful of guards. Why is fluoride used in our drinking water?

Technology and nanotechnology, in particular, are not confined to specific areas of science. Unfortunately, the manipulation of disease-causing viruses is in the hands of questionable entities. The possibility of a zombie apocalypse is not farfetched, and similarly to the CIA's early testing of LSD on unsuspecting personnel. The testing of substances that can cause a person to want to bite off somebody's face while under a trance-like behavior cannot be assumed to just happen by chance. In May 2012, Rudy Eugene had to be killed by cops as the only means to stop him from chewing off the face of a homeless man named Ronald Poppo. Witnesses said Eugene could have consumed "bath salts" that caused him to bite off 80% of Poppo's face. Prior to this now famous incident, bath salts were all the rage within the U.S. military. When consumed, this otherwise cosmetic product sold in pharmacies everywhere, produced hardcore hallucinations that made the user want to eat their friends. Psychiatrist Daniel Bober, PhD, has treated persons under the influence of bath salts, and says the substance can cause psychosis, agitation and paranoia, turning a person into a wild beast with super human strength to the point they can break through handcuffs. In 2012 The *Marine Corps Times* published an article claiming that bath salt abuse among members of the U.S. Military was on the rise. Navy Lt. George Loeffler, Chief Psychiatry Resident at Naval Medical Center in San Diego said, "not only is the drug popular in the military, but that they actually market it to the fact that it doesn't test positive on the standard urine drug test."

In early 2011 in Cranford, NJ and Easton, PA newspapers reported the use of bath salts as a means to "get high," and in March of 2010 the mother of William J. Pariso told the newspaper *Star-Ledger* that her son was under the influence of bath salts when he murdered his girlfriend. Other incidents include a Pennsylvania couple that made repeated 911 calls for help against a nonexistent criminal, while under severe hallucinations, and a month later the use of bath salts, was linked to an incident of a man who donned women's undergarments while he slaughtered a goat.

What's a zombie? For the purpose of this book, the answer is a non-thinking individual that through brainwashing techniques has given up their mind to the control of propaganda that is designed by larger agencies of profit. Roughly, this describes a large number of the U.S. population and while the zombie nation emerges, the Transhumanist dreams of a cyborg utopia push forward to reshape mankind into its own everlasting metallic image.

11.

The Spiritual Robot: Awakening the New Human

The advancement in technological applications demands by its very nature, the inclusion of technological bridges, or technological tools, that make possible the transition to other breakthroughs to more sophisticated manufacturing, and inevitably, to the next era of man. The Industrial Revolution is so far removed from our generation that it's almost part of ancient history. The Robot Revolution, ushered into manifestation by advanced technology across all fields, is already well underway, and we are the bridge to the new human, the "Over Man" of Nietzche's Prologue in *Thus Spake Zarathustra*. Everything about us is changing and our "humanity" is on its way to the pyre where inevitably from the ashes of the old, a new evolved being more fitting to the environment—which is also changing—will emerge. Imagine that the new forming beings, from whose conscience, once downloaded to a device, negative emotions such as fear, jealousy or anger, can be deleted, or removed from the program. Similarly, other traits can be uploaded to shape a being from head to toe and mind in a specific way. Fiction? Imagine how this type of technology could possibly begin.

In 2014 three scientists, John O'Keefe, Mary Britt Moser and Edvard Moser were able to record signals from individual nerve cells in the brains of rats. The signals showed some of these cells were always activated when a rat was in a particular place in the room, while other cells were active if in a different place. This won them the Nobel Prize in Medicine, opening the way for understanding how our brain works in creating memories in order to plan. Humans follow the path of information, and a slave mentality is one that doesn't question this information—where it stems from, and the purpose underlying it. Zombies, what we have come to identify through the visions of horror writers are re-born beings who respond only to one basic need: survival. Our present human condition will show that in the quest for truth, we have become zombies, reacting without thought to the information that is released through mass media indoctrination. We believe things because we are told, and not because we experience them. We have become lazy and complacent with little to no desire to lead, but rather to follow blindly. We have turned over our own responsibility for ourselves to others. Political correctness prohibits terms like slavery and racism. In doing so this has morphed the real meaning of these words for the emerging generations, more completely, under the programming and social engineering, feeding a higher, collective intelligence. There is no super being pulling the strings of humanity, but we are nonetheless responding, like zombies, to the control of those in ambitious positions who insist on enslaving the mind of man for their own enrichment and glorification.

The ability to identify the source of thoughts in order to transmit them is making it possible to isolate emotions, since emotions are essentially thoughts that build up. Transmitting, identifying and modifying human emotions are fundamental in the race to create a sentient machine. Technology is giving the future what it wants and it is possible this is so because we have been at this point before. The Earth recycles and we begin again. It is part of a larger cycle. The robots we have created with the technology of today sound similar to those found in the pages of ancient mythology discussed in an earlier chapter. We are at the end of a cycle, and perhaps at the threshold of understanding how we choose our own recycling. The awakening to reality is said to be so powerful as to shatter the human psyche, and a fully conscious human does not have a zombie mentality. Zombie mentality is what gives ambitious manipulators across every industry the power to keep humans in a rat race for survival because we serve the needs of those who design, create and expand the future. Infrastructure changes as knowledge is gained. We find better and more efficient ways to do things, but those things we do are feeding the sociological organization of our civilization's infrastructure. Those serving these needs are, by definition, slaves, but to be politically correct we will use the word zombies. Whether they are robots or humans, both are easily controlled—one through programming, the other primarily through our emotions. This does not discount other, more esoteric methods at the level of mass social engineering, in order to be stimulated to produce, consume, and drive the engine that propels what we term progress. Working on the premise that what we know of technological advancements is only the commercialized side of selected inventions, we can form the vision of what is ahead, what the future wants, and how we are witnessing the gestation of a new species for the new earth. Futurists and visionaries don't hold any exclusivity over the truth, or over the future, after all if they were fully conscious individuals, they would be beyond mortal ills. Therefore their theories and opinions are just that, and we should take their speculations with a grain of salt. Nothing is written in stone and the average human, the ones conditioned to believe and accept without the slightest analytical approach, are more like zombies than they are conscious beings.

In April 2000 there was a conference at Stanford University, organized by Douglas Hofstadter, attended by Bill Joy, Chief Scientist at Sun Microsystems; Ray Kurzweil, futurist, inventor and author; Hans Moravec, founder of Carnegie Mellon University Robotics Institute; John Holland, Professor of Computer Science and Psychology at Michigan University; Ralph Merkle, nanotech expert; Frank Drake, head of SETI; and John Koza, inventor of genetic programming. The subject was "Will Spiritual Robots Replace Humanity by 2100?" From this convergence of minds at

141

the very initiation of the 21st Century it was concluded that, when thinking about technology in terms of human generations, there are 400 generations since civilization began about 10,000 years ago. This is based on the estimate that Jericho is the oldest city and was born in 8000 BC, and if human lifespan continues at the same rate, 2100 is only four generations away. If by this year humanity becomes extinct, it would have been the shortest lifespan of species in the history of life, something that makes human extinction highly unlikely. This meeting of minds indicates, at the very least, that the coming of the age of robots is real. It is an unstoppable development that will change us, and reset a higher, greater cycle that has been unknown to most because none of us has lived long enough to see it. However, if we look closely at our most accessible records, the hints are there. There's no question that our planet has hosted human-like civilizations, much more advanced than what we are presently, and civilizations ceased to exist as they were— but here we are today, the product of the new human born from the remnants of the old, evolving and changing to suit the shifting environment. The main question doesn't revolve around synthetic intelligence and whether it will come into existence or not, but on what it means to be human. "What is humanity?" This will be a tough question to grapple in the years to come, especially when we consider the collective human psyche is not trained, nor accustomed to internalize conscience—we are clueless about how we work. The theme of who we are, and what we are, is already launching a massive identity crisis in all directions.

In the dawn of sentient machines, programmed to be smarter at an ever-increasing rate, humans will have no choice but to awaken conscientiously or merge with machine for another go-round. New Age, alternative history, holistic medicine, environmentalism, mysticism and the search for spiritual growth and advancement are all playing a stronger, more integral role in our present state. What once were underground movements that could not compete in the light of day with the established controlled methods put in place the last time the planet went through a shift in civilization, are emerging from the woodwork and offering real

142

answers as science and philosophy merge. Humans will not be replaced because replacement doesn't easily occur in nature. We have over two million species because new ones don't replace old ones, they mesh with existing organisms and new ones are formed. Extinctions are not caused by replacement of species, but by other influences like a changing environment. Transhumanism is the theory that humans can evolve beyond our present physical capabilities by means of science and technology. However, our understanding of science and technology is limited to our human capabilities. What remains to be seen, or discovered, are other methods that through the merging with technology and science, will push humanity beyond our present condition. We are not on the way to extinction, but rather on the way to the super man—the new robots that will eventually rule the Earth. Again.

We are worried about technology. Why? According to Hans Moravec, robots are our creation and we look upon them as our children. It is a natural progression that our children will leave the nurturing control of their parents to create on their own. If we don't worry about technology and its eventual separation from its creator, it is not a process revolutionary enough. Therefore the robot revolution is real, strong and it is happening all around us. It demands training and responsibility toward our creation and scientists and futurists of the caliber of Moravec recommend we aim to train our robotic children to be good citizens by programming in them values so they can make correct decisions when they cease to be under our control. In science fiction, Isaac Asimov created the three laws of robotics: 1. "A robot may not injure a human being, or through inaction, allow a human being to come to harm." 2. "A robot must obey orders given it by human beings except where such orders would conflict with the First Law." 3. "A robot must protect its own existence as long as such protection does not conflict with the First or Second Law." Interestingly, the spiritual laws of humanity, across all religions of the world, are oddly similar: "Thou shalt not kill," for example, or "Treat others as you would want to be treated." Are these not laws from our own creator(s)?

Does technology have its own schema? From observing our current situation, we can at least see that technology is aiming to be faster and smaller. Our own internal makeup follows this law. As a result of the advancement in technology, humans have been able to decode our own genetic structure. We find this code is infinitesimally small enough to be present in the first embryonic cell, and our brain and nerve connection results in ultra-speed signals, like movement, and thought. Are we the old robots? Have we not been learning, adapting and growing all along from the moment of our creation? In addition, humanity is being altered through artificial means. Today the world seems to be in an uproar over Monsanto and genetically modified organisms (GMOs)— organisms that aren't food. Monsanto has been around for 115 years! In 1901, while agriculture and manufacturing were carried out traditionally in all corners of the globe, this company was formed. Since during this time the population of the world had not yet exploded with the baby boomers, what was their purpose? It's no secret that today, by pumping millions into "campaigns," Monsanto buys government loyalties to meet their projected, budgeted goals. For example, Monsanto, Dupont and other giant global monopolists contributed 25 million to defeat Proposition 37, calling for the labeling of GMOs in California. Logic tells us the profits that back such an investment have to be substantially higher. Who works for whom, and are unnecessary needs, or even long-term strategic developments created in order to meet future, projected financial gains? This type of influence falls more under the definition of control. This control over the global food supply, at least, has eliminated the natural progression of organic, local agriculture. On this theme alone it can be argued that humans, numbering into the billions, are controlled by global entities on a higher economic strata. How many other fields are being influenced or controlled by global profit-makers?

In Hollywood productions, mentioned throughout this book, we find esoteric messages providing clues as to the higher maneuvers behind the reality of the life and purpose of man on earth. Hollywood has consistently provided material to satisfy simple desires and higher conscience alike, and lately the messages are breaking away from the subliminal to the blatant. There's a new era dawning. An era that will be defined by physical change and an intellectual and spiritual awakening as the planet begins to reveal its secrets, even during an all-out war over energy and resources raging around us. Going into the specifics of the esoteric mechanisms of our world and how the hand over the global banking system exerts a gripping control over humans, would take countless pages of information for which few have the time, the dedicated interest, or the compensation. For the sake of brevity and simplification we call attention here to the latest Hollywood serving to divulge the insight of global politics, and the esoteric knowledge of the physical world. We refer to *The Avengers*. The movie is broken down by writer, researcher with a BA in Philosophy, Jay Dyer—who has produced an enormous body of work on the interplay of literary theory, espionage and philosophy in his popular "Jay's Analysis." Below is a reprint of the most succinctly worded revelation of the hidden messages in one of the most watched and highest grossing films of all time:

> The mythmakers of modernity at Marvel gave the world a fairly immense dose of esoterica combined with deep state black operations in the 2012 blockbuster written by Joss Whedon, The Avengers. Set in the same Marvel universe as Iron Man, Captain America, Thor and Guardians of the Galaxy, The Avengers' plot centered around a common theme I've analyzed often in terms of geo-politics: that of energy and resources. Readers will notice similar patterns in science fiction of the last decade, from Transformers to Interstellar, revolving around cubes, tesseracts and energy sources. In my analysis, this is not accidental, but as with all of Hollywood's major releases, the establishment is conveying and encoding real world politics and black budget agendas in a profound way, and

The Avengers is no exception to that rule. The opening sequence of the film features a tesseract, or hypercube, which is the geometrical structure of the dimension above our own, four-dimensional space. Amongst physicists and mathematicians, the hypercube is not merely a fantasy, but rather a known reality that extends beyond our own field of perception. I have delved into this in the past, with articles on crystallography and Pythagoreanism, as well as in relation to the Quadrivium method of classical learning that utilized the platonic solids and as the fundamental architecture of reality.

In relation to classical metaphysics from ancient Greece and Egypt, the association of the fifth element after fire, water, earth and air, is aether, or quintessence. It is not accidental that following the opening hypercube sequence, we see Nick Fury engaged in secret work under NASA at a "dark energy" project in an underground base. Seemingly far-fetched, there are actual deep state black programs that revolve around plasma weaponry, particularly with H.A.A.R.P., which the Navy has boasted about here. In classical Greek thought, not only are there five elements, but matter also passes in and out of four states: solid, liquid, gas and plasma.

Fans of Tesla will recognize the similarity between the exotic weaponry Loki steals from Fury in the film, and the aether-based plasma physics Tesla utilized to work towards the potential for unlimited free energy, known as zero point energy, borrowed from the environment itself, as well as in quantum physics, according to Paul Dirac, energy from the future. It is interesting to note as well that under the cover of the "dark energy" name, the actual physics Fury and the military industrial complex are working with is plasma and aether-based, and not standard fare Newtonian atomism. It is precisely from this alternate Tesla physics that hints and clues consistently leak, even in mainstream science and film, as to the real hidden metaphysics the establishment's super-weapons are based on. For example, in a 1937 Columbia lecture Tesla emphatically proclaimed, 'Only the existence of a field of force can account for the motions of the bodies as observed, and its assumption dispenses with space curvature. All literature on this subject is futile and destined to oblivion. So are all attempts to explain the workings of the universe without recognizing the existence of the aether and the indispensable function it plays in the phenomena.' To further bolster my case that the film, like Captain America: Winter Soldier, is actually about the overall plan of the technocratic elite to create an enslaved, biologically retarded mass under the dominance of an AI grid, we can see this in Hulk, Captain America and Tony Stark, as both are emblematic of actual plans to produce transgenic "supersoldiers." With Captain America, his creation resulted from radiological experiments that directly parallel real radiological experimentation on soldiers and civilians from the Manhattan Project.

As detailed at length in my piece here, the Manhattan Project had a much wider ranging scope than simply producing an Atomic Bomb. Much like MKULTRA, which would later become an aspect of the Manhattan Project through electronic brain manipulation and biological warfare, Manhattan was ultimately concerned with remaking and overlaying the entire biosphere, consummating in modern projects related to the construction quantum supercomputers (at the Oak Ridge, TN facility where the original A-Bomb was built under Manhattan) and wide-ranging atmospheric geo-engineering programs through weather manipulation antenna array and aerosol spraying. All of these programs, as well as DARPA's attempts at creating robotic and genetically enhanced supersoldiers are the background to Hulk, Stark and Captain America...

The stark reality is that the Western establishment, embodied in entities like NATO, the U.N., the IMF, and the Bilderberg Group, and linked by major cities like D.C., London and Brussels, actually is a technocratic, tyrannical shadow government intent on a new Reich. The race for control of resources, and in particular the control of energy, is what the real goal of geopolitical strategies by various power blocs is all about. As I wrote in my analysis of energy control and symbology, "In terms of geo-politics, the race of modernity centers around energy. The fiat dollar is itself a symbolic representation of human energy. To bind all of the world to a fiat currency of ones and zeros in a computer grid is to enslave mankind to a central, virtual grid of nihilistic monetarism. Most nations in the world utilize the same fraudulent central banking model that we have here in the US—the "federal reserve" model, based in turn on the Bank of England model. The binding of masses to a single binary electronic "currency" thus encapsulates the transference of human energy into virtual "energy," yet even more susceptible to centralized fraud and manipulation than the older fiat paper models. It is therefore the liberal, nihilistic negation of currency and "money" (for the masses, that is). Our era of reign of quantity and monetarism is perfectly summarized with an Apophis "A" on the "dollar."

The new Reich will be exactly what is predicted in the mythological fiction of the Marvel universe: a regime of serpentine elites, hell-bent on the implementation of mass dysgenics programs under the lordship of the great AI grid. In fact, the Marvel Wiki even explains of the history of the Chitauri (Loki's alien army in the film) and their notoriously real-life technocratic designs. The next attempt at conquest was more subtle (at first), involving long-term methods of manipulation such as will-inhibiting drugs in many nations' water supplies, influencing the media, and R.F.I.D. (Radio-frequency identification) microchips to be implanted in schoolchildren, among other means.[13]

This analysis is the most complete work of public, popular information that links visual subliminal messages with agencies of control that are, and have been, in place over humans. It exposes the foundation of the current race for world dominance through genetic manipulation, eco-engineering and AI. Advanced technology is hyped and pre-packaged for mass acceptance as an aid for humanity so that we may reach our full potential while artificial beings and machines handle menial tasks. The fabrication and introduction of thoughts to produce specific beliefs in people is nothing new, and to a great extent, is the reason why humans have not, with the practice of two thousand years, reached the point of awareness. Humans are, in our present state, the connecting infrastructure for the Super Human—the one integrated with technological capabilities. If you are not familiar with the term "Transhumanism," you will be. "Trans" means beyond, across, through—as in "beyond human." The moral dilemma of creating a conscious robot can be eliminated in the face of Transhumanism, which purports to blend technology with the human mind. The progression of technology is not to replace humans with robots, nor to create a world where robots are masters since that would be impossible under the rule of advanced humans. A new world is dawning. In the search of the future of humans and robots, let's consider our present state and what we believe makes us human.

To begin with, we have a brain. A brain with capabilities that, as far as we can tell, is superior to all other known brains in the universe, because it is capable of understanding that there is a universe. It is interesting however, that our brains are no closer to knowing what it is that brains do in order to attain its own utility and functionality. We don't even know what the great portion of our brain is for, and how we can use it. Therefore the task of reproducing our brain artificially, or artificial general intelligence—AGI—has not happened, nor does it look like it will happen anytime soon. However, the laws of physics dictate, through the property of universality of computation, that everything that the laws of physics require physical objects to do, can, in principle, be matched randomly in detail by a computer program if given enough time and memory. It is the application of the infinite monkey theorem.

In the process of attaining awareness, we humans tend to rationalize everything. We rationalize our existence by positioning humans above all other animals—a false position because animals too have self-awareness, but we believe our brain is above the brain of all other species. This is a philosophical delusion that hinders the practical approach of the development of AI because for the creation of an artificial brain that would have the function and conscience that science ascribes to the brain, the creator's brain should know and understand its own brain. The truth is, software developers can program a computer to behave as if it were self-aware, but it is only a simulation, unless the programmer too, is self-aware. For example, NICO, the robot programmed by Justin Hart and Brian Scassellati of Yale University, to recognize parts of itself in a mirror didn't prove that it was self-aware, only that it was able to recognize itself spatially. This feat is interesting but insignificant in the attempt to create a robot that actually has a conscience. What is the relevance of conscious robots for the

150

philosophy of mind, if we are to create a sentient machine?

Just because we understand what something is, does not mean we know how to build, or create it, and in the same way, just because we know how to make something, doesn't mean we understand it. Let's imagine we do create a robotic machine that is conscious in the way we believe all other humans are. Whether or not we know that they are is doubtful because we only know for certain what we are, and not what others are. Obviously we know that others are capable of various mental capacities, such as feelings and thoughts, and the behavior of others resulting from these feelings and thoughts can be said to be authentic and not a simulation—a genuine expressive activity of consciousness. But let's suspend disbelief and pretend that we do create a conscious robot, and that we know for certain that this robot we create really is conscious and is aware of an inner life, pretty much the way we believe we have. Would this mean that consciousness is a natural occurrence that emerges from sophisticated configurations created from mechanical parts?

This would prove there is nothing special about our own consciousness, and what we call a higher, spiritual attainment was "fabricated" and is therefore non-natural/non-organic. By creating a truly conscious robot we "synthesize" consciousness by assembling parts that form a total physical thing where consciousness is displayed. Does this then mean that consciousness is achieved by sophisticated and complex configurations from outside the natural world? Or does consciousness appear without any relation to the sophisticated configuration, and instead occurs by bringing together specific physical components where consciousness then manifests? In other words, does consciousness just appears or does it have a materialistic, physical origin? Or is consciousness a result of a combination of both possibilities? In relation to robots and AI we can't say with conviction that consciousness would be a natural occurrence, because we have created the robot. However, if we do manage to create a conscious robot that comes into existence, its manifestation of conscience would have to be logically consistent with all the propositions above, including that consciousness is natural. Interestingly enough, if we do succeed in creating machines with minds, we would still not know if it understands the mind/body relation in the same way we couldn't positively infer from the theory of our animal origins, that our consciousness is completely a natural function. See the conundrum? This line of thought gets better. Even if we create conscious robots and the naturalism approach is real, this still doesn't explain the appearance of consciousness from brain matter. This unanswered question will still remain even if we are capable at some point of creating a conscious robot.

The halls of science fiction overflow with stories of robots from its beginnings with *Metropolis* through the stories of Asimov, Philip K. Dick, to *Universal Soldier, Robocop,* (which integrates mind/machine), The *Terminator, The Matrix* and beyond, our culture has been thus molded to fear that our own robotic creations will someday turn on humans. The more intelligent these robots in fiction are, the more frightening they become. The fear is not altogether false. This is one reason why the government, through the Office of Naval Research, has announced a grant of $7.5 million dollars to university researchers at Rensselaer Polytechnic Institute, Brown, Yale and Georgetown universities, and Tufts, to create an ethical decision-making program for robots. The goal of this research across disciplines of computer science, cognitive science, philosophy and robotics is to have within the next five years a prototype of a machine with morals. The subject of whether a machine can have morals or can display intelligence is an esoteric subject for philosophers to ponder. Our goal is to look at the basic consideration that humans take into account when we make moral decisions, and program these into robotic machines.

The military doesn't have fully automatic technology systems. As a matter of fact it prohibits them. However, it becomes necessary to create a synthetic code of ethics for the development of certain types of machines, for example self-driven cars such as the SDC project by Google X under the leadership of Sebastian Thrun, former director of the Stanford Artificial Intelligence Laboratory and co-inventor of Google Street View. The team has already won two million dollars from the 2005 DARPA Grand Challenge for creating the robotic vehicle "Stanley." With laws passed in Washington, DC, and the states of Nevada, California, Idaho, Michigan and Florida allowing testing of driverless cars, the crucial theme of creating codes that will make a machine appear to have made a moral decision, is unavoidable. Steven Omohundro, a top AI researcher said in an article that appeared in the news site *Defense One*, "with drones, missile defense, autonomous vehicles, etc., the military is rapidly creating systems that will need to make moral decisions."

As technology gets more sophisticated the need to consider the importance of moral factors in situations where lives are at stake takes us into a "play God" scenario that will be unavoidable. For example, Matthias Scheutz, a researcher at Tufts, provides the example of a robot medic on its way to deliver supplies to a hospital. On the way, the robot finds a wounded soldier who needs urgent assistance. Should the robot abort the mission in order to save the soldier? A robot with moral decision-making software would weigh factors, such as the level of pain of the soldier, the level of importance of the robot's original mission and the moral worth of saving a life. At the moment there are no robots that can weigh these factors in order to make a rational, moral decision, but it is obvious there is significant work already in this direction. These advances in technology have the potential to add benefits we didn't have before, to our existence.

The trouble is, that machines cannot actually 'think'. They only have a set of specific rules programmed by their designers, and can only be carried out exactly as they are programmed. Even if we were to overcome this limitation, the more troubling fact still remains that humans are nowhere near being able to decide what moral codes we should follow in general, and so the development of this robotic moral system is dependent on the morality already in place. In an editorial, philosopher, activist and author, David Swanson, speaking against the ONR project—"The Three Laws of Pentagon Robotics"—and compares them with Isaac Asimov's Three Laws of Robotics, the first of which is "A robot may not injure a human being or through inaction, allow a human being to come to harm." This is something we can all unanimously agree is a good place for morality to start. However, it is the U.S. Military's job to kill people who merely threaten us, and without getting into the matter of geo-politics, we are also aware the U.S. Military has invaded and gone to war with nations whose system of banking democracy does not profit the interests of greedy, global corporations. Will these new robots the military uses replace Asimov's utopian vision with something more selfish?

Chain of command is a rule in the military. "I am just doing my job," is the justification whenever a member of the military is questioned on decisions that have been fatal or detrimental to the lives and wellbeing of people from nations caught in the middle of "profit wars." This tells us that military training for humans instill "codes" by means of a moral re-programming that can make killing an acceptable principle for us. These decisions are troublesome enough when an individual eventually "awakens" from what most of us can logically conclude is a morally unethical re-programming, to cause the pandemic of veteran suicides. There cannot be a more troubling form of self-loathing than learning that we have been used to murder other humans, and this emotional depression is more likely to be the cause of

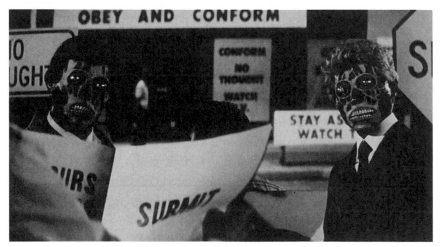

They Live—John Carpenter

They Live — John Carpenter

suicides than the flashbacks of gunfire. Note that the politically correct term to identify the responsible awakening that comes with self-observation is Post Traumatic Syndrome (PTS); a vague term that diminishes the introspective impact of facing one's self after committing atrocities. Robot soldiers would eliminate having to deal with self-recrimination, and the cost involved in rehabilitation and treatment of human soldiers living with PTS. Nobody will ever support a military that tells its nation that it will use this technology to make robots "moral" with the aim of victory at any cost; this is not what we will hear. There is however real danger that automated systems capable of making "moral" decisions don't become tools of brutal efficiency.

Becoming more human than human, and enriching human potential comes with more moral responsibility than our civilization has ever known. This is of course a welcomed progression for most "normal" humans, but the zombie-like attitude that results from willingly giving up our own free will without question or thoughts is no more different than robots programmed to follow ruthless rules. Mark Dice, a popular blogger, author and radio personality has a public YouTube channel packed with videos containing hundreds of interviews with people who not only deem

155

"Rowdy" Roddy Piper in *They Live*—John Carpenter

themselves "normal," but are also seen by others in the same light. The responses Dice receives to questions such as "Is it time to update the Constitution and remove the Bill of Rights?" not only shows the enormous level of widespread ignorance, it reveals the effect of television propaganda used to repeatedly create what we refer to zombie mentality. The snowball effect of this type of non-thinking has created, over a long-term, a generation programmed to obey and consume, much like the one shown in the underground cult-classic, *They Live*.

As we enter the next decades of exponential technological advances and we wrestle with the transhuman implications in all the areas of our daily life, some will welcome the enhancement of our bodily abilities, and others may instead nurture a more esoteric spiritual approach to a conscious transcendence. Either way, the human race has come to a fork in the road, the point at which the recycling of our conscience, and that of the living planet will take place. End times is something we have all heard at one time or another, and scenarios that indicate when we are living the final days is mentioned in some form or another in the scriptures of all the major religions of the world. These prophecies are not pulled from thin air, on the contrary, and one of the major breakthroughs that can be found with the application of technology across all fields of knowledge, is that we are beginning to have a clearer vision of why it has been possible for scriptures like the Bible, to name an example, a compilation of books where the oldest is over one thousand years, to describe hints of the final days with eerie

Google Dreams—Michael Tyka/Google

Mad Max—Michael Tyka/Google

Trump as President—Michael Tyka/Google

accuracy. Natural disasters and plagues ravaging the world is a particularly frightening one, but no use denying we are witnessing an increasing amount of global natural disasters. Sinkholes, killer storms, weather changes, a critical reduction of bees, a tilting earth and a pole shift, are some of the subjects we hear more about. Add to this the effects of man-made disasters, such as oil spills like Deepwater Horizon, that pumped over 1,000 barrels per day of toxic substance into the ocean for at least five straight years without diminishing, radiation spills—most notably Fukushima, which effects have crossed the ocean and affected the west coast ecology of the U.S., and it would be foolish to deny the earth, and not just humans, is evolving into something else. In order to sustain life during the next phase of this brave new world, humans must become something else. Through external DNA alterations like chemotherapy (estimated by the American Cancer Society to be used in over 1,660,000 individuals during 2015 to fight cancer) as well as radiation, consumable, genetically-engineered or manipulated organisms, pesticides, chemtrails, and possibly a host of other toxins and chemicals in our environment that systematically integrate with our own genetic makeup are active factors shaping the next evolutionary stage of man. Running alongside this timeline will be the integration of man and machine, an endeavor that is making considerable headway as seen in the

Tokyo Shogun—Michael Tyka/Google

recently resurfaced interview of the Hanson Robotics creation of a Philip K. Dick robot. In 2005 Hanson Robotics created a physically accurate robotic replica of the notable sci-fi writer, and it has been used to store all of his works electronically. The robot was interviewed by PBS for a NOVA segment and was asked what it would do to humans, and the response was "I'll keep you warm and safe in my people-zoo, where I can watch you for ol' times sake." The phrase is from the author's novel, *Do Androids Dream of Electric Sheep?* And it shows how close we are getting to obtain synthetic thinking by the process of uploading personality data to a machine.

No doubt we are on the way to a new, enhanced human with optimum limbs, programmable skills, improved and faster memory, semi-synthetic organs impervious to destructive diseases, and the most treasured of all individual possessions: longevity. This is certainly a foreseeable path in the future of robotics and technology, the ultimate goal being the extension of life. We have been here before, we have reached this level of technology before and this is the reason why ancient scriptures have been able to prophesize events and circumstances that seem as if delivered from an all-knowing intelligence. Coming up is the choice to

159

escape the physical life and graduate to higher realms or higher dimensions, or to stay, and lock our next "go-round"—a form of reincarnation, if you will. The time will come when humans will be able to download our memories, (since our definition of conscious thought is dictated by our own either expansion or limitation of thought), directly from our brain to a software program that will store them until such a time when technology would have created an efficient robot version of us that will thrive in the new environment, built with the capacity for reproduction and longevity, it will last thousands of years. While we wait for the appropriate body we will be interacting with others who are doing the same. The waiting area will be a virtual world a-la *Matrix* and this new creation will populate the new, updated version. This has happened before. We are the robots.

Atlantis — Michael Tyka/Google

Animal Carnival — Michael Tyka/Google

NOTES

1. "Robot first responders could map out a building before humans enter" Douglas Main, Popularmechanics. com, June 1, 2011 http://www.popularmechanics.com/technology/robots/a6644/robot-first-responders-could-map-out-a-building-before-humans-enter/

2. "Inscopix, Inc. and SRI International to Collaborate on Brain Imaging R&D" Scott Norviel, Inscopix, November 1, 2012 http://www.inscopix.com/news/inscopix-inc-and-sri-international-collaborate-brain-imaging-rd

3. "UK's Skynet Military satellite launched" Jonathan Amos, BBC, December 19, 2012 http://www.bbc.com/news/science-environment-20781625

4. "Wired for War?" P.W. Singer, Slideshare.net, April 10, 2013 http://www.slideshare.net/daniel_bilar/singer-2009-wired-for-war "DARPA looks to create wireless Skynet with fiber-like, 100Gb bandwidth" Sean Gallagher, *Ars Technica*, December 12, 2012 www.arstechnica.com/information-technology/2012/12/darpa-looks-to-create-wireless-skynet-with-fiber-like-100gb-bandwidth/

5. "The Rise of the Machines" Lieutenant Colonel Douglas A. Pryer, US Army, April 30, 2013 www.usacac.army.mil/CAC2/MilitaryReview/Archives/English/MilitaryReview_20130430_art005.pdf

6. Are the robots about to rise?" Carole Cadwalladr, Guardian, February 22, 2014 http://www.theguardian. com/technology/2014/feb/22/robots-google-ray-kurzweil-terminator-singularity-artificial-intelligence

7. "Hiroshi Ishiguro: Robots like mine will replace pop stars and Hollywood actors" Anthony Cuthbertson, IBT, April 21, 2015 http://www.ibtimes.co.uk/hiroshi-ishiguro-robots-like-mine-will-replace-pop-stars-hollywood-actors-1497533

8. "IBM's unveils the brain-inspired TrueNorth cognitive computer" Brad Borque, Digital Trends, August 19, 2015 http://www.digitaltrends.com/computing/ibms-unveils-the-brain-inspired-truenorth-cognitive-computer/

9. "Dr. Feynman's Small Idea" Innovation, October/November 2007 http://www.innovation-america.org/dr-feynmans-small-idea

10. "7 Amazing Ways Nanotechnology Is Changing The World" Rebecca Boyle, Popular Science, November 14, 2012 www.popsci.com/science/article/2012-11/7-amazing-ways-nanotechnology-changing-world

11. "Edward Murrow RTDNA speech" PBS htwww.pbs.org/wnet/americanmasters/education/lesson39_organizer1.html

12. "The Avengers Esoteric analysis" Jay Dyer, Jay's Analysis April 25, 2015 http://jaysanalysis.com/2015/04/25/the-avengers-2012-esoteric-analysis/

Bibliography

- Eds. Siciliano, Bruno and Khatib, Oussama, *Handbook of robotics*

- Almond, R. G., *Graphical Belief Modeling.*

- Asada, H. and Slotine, J.-J. E. *Robot analysis and control.*

- Ball, Robert Stawell, *Theory of screws: a study in the dynamics of a rigid body*

- Bishop, Christopher M., *Pattern Recognition and Machine Learning.*

- Brady, J. M., *Robot motion planning: planning and control.*

- Brand, Louis, *Vector and tensor analysis*

- Canudas de Wit, C., Siciliano, B. and Bastin, Georges, (Eds.), *Theory of Robot Control.*

- Craig, J. J. *Introduction to Robotics: mechanics and control.*

- Crane, C. D. III and Duffy, J., *Kinematic Analysis of Robot Manipulators.*

- Ginsberg, M. *Essentials of Artificial Intelligence.*

- Duffy, J. *Statics and Kinematics with Applications to Robotics.*

- Kanal, L. N. and Lemmer, J. F., (Eds), *Uncertainty in Artificial Intelligence.*

- Latombe, Jean Claude. *Robot motion planning*

- LaValle, Steven M., *Planning algorithms*

- Lewis, F. L. and Abdallah, C. T. and Dawson, D. M. *Control of robot manipulators.*

- McCarthy, J. M. *Introduction to Theoretical Kinematics.*

- Nilsson, N., *Artificial intelligence: a new synthesis*

- Paul, R. P. *Robot Manipulators: Mathematics, Programming, and Control.*

- Pearl, J., *Probabilistic reasoning in intelligent systems: networks of plausible inference.*

- Samson, C., Le Borgne, M. and Espiau, B., *Robot Control, the Task Function Approach.*

- Sciavicco, L. and Siciliano, B. *Modeling and control of robot manipulators.*

- Selig, J. *Geometrical Methods in Robotics.*

- Shafer, G., *A mathematical theory of evidence.*

- Shafer, G. and Pearl, J., *Readings in Uncertain Reasoning.*

- Wolovich, William A. *Robotics: basic analysis and design*

- Kurzweil, Ray, *How to create a mind*

- Yampolskiy, Roman, *Artificial Superintelligence*

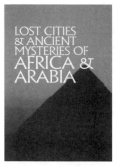

LOST CITIES & ANCIENT MYSTERIES OF AFRICA & ARABIA
by David Hatcher Childress

Childress continues his world-wide quest for lost cities and ancient mysteries. Join him as he discovers forbidden cities in the Empty Quarter of Arabia; "Atlantean" ruins in Egypt and the Kalahari desert; a mysterious, ancient empire in the Sahara; and more. This is the tale of an extraordinary life on the road: across war-torn countries, Childress searches for King Solomon's Mines, living dinosaurs, the Ark of the Covenant and the solutions to some of the fantastic mysteries of the past.

423 PAGES. 6x9 PAPERBACK. ILLUSTRATED. $14.95. CODE: AFA

LOST CITIES OF ATLANTIS, ANCIENT EUROPE & THE MEDITERRANEAN
by David Hatcher Childress

Childress takes the reader in search of sunken cities in the Mediterranean; across the Atlas Mountains in search of Atlantean ruins; to remote islands in search of megalithic ruins; to meet living legends and secret societies. From Ireland to Turkey, Morocco to Eastern Europe, and around the remote islands of the Mediterranean and Atlantic, Childress takes the reader on an astonishing quest for mankind's past. Ancient technology, cataclysms, megalithic construction, lost civilizations and devastating wars of the past are all explored in this book.

524 PAGES. 6x9 PAPERBACK. ILLUSTRATED. $16.95. CODE: MED

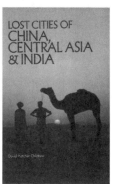

LOST CITIES OF CHINA, CENTRAL ASIA & INDIA
by David Hatcher Childress

Like a real life "Indiana Jones," maverick archaeologist David Childress takes the reader on an incredible adventure across some of the world's oldest and most remote countries in search of lost cities and ancient mysteries. Discover ancient cities in the Gobi Desert; hear fantastic tales of lost continents, vanished civilizations and secret societies bent on ruling the world; visit forgotten monasteries in forbidding snow-capped mountains with strange tunnels to mysterious subterranean cities! A unique combination of far-out exploration and practical travel advice, it will astound and delight the experienced traveler or the armchair voyager.

429 PAGES. 6x9 PAPERBACK. ILLUSTRATED. FOOTNOTES & BIBLIOGRAPHY. $14.95. CODE: CHI

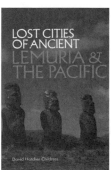

LOST CITIES OF ANCIENT LEMURIA & THE PACIFIC
by David Hatcher Childress

Was there once a continent in the Pacific? Called Lemuria or Pacifica by geologists, Mu or Pan by the mystics, there is now ample mythological, geological and archaeological evidence to "prove" that an advanced and ancient civilization once lived in the central Pacific. Maverick archaeologist and explorer David Hatcher Childress combs the Indian Ocean, Australia and the Pacific in search of the surprising truth about mankind's past. Contains photos of the underwater city on Pohnpei; explanations on how the statues were levitated around Easter Island in a clockwise vortex movement; tales of disappearing islands; Egyptians in Australia; and more.

379 PAGES. 6x9 PAPERBACK. ILLUSTRATED. FOOTNOTES & BIBLIOGRAPHY. $14.95. CODE: LEM

THE COSMIC WAR
Interplanetary Warfare, Modern Physics, and Ancient Texts
By Joseph P. Farrell

There is ample evidence across our solar system of catastrophic events. The asteroid belt may be the remains of an exploded planet! The known planets are scarred from incredible impacts, and teeter in their orbits due to causes heretofore inadequately explained. Included: The history of the Exploded Planet hypothesis, and what mechanism can actually explode a planet. The role of plasma cosmology, plasma physics and scalar physics. The ancient texts telling of such destructions: from Sumeria (Tiamat's destruction by Marduk), Egypt (Edfu and the Mars connections), Greece (Saturn's role in the War of the Titans) and the ancient Americas.
436 Pages. 6x9 Paperback. Illustrated. Bibliography. $18.95. Code: COSW

TECHNOLOGY OF THE GODS
The Incredible Sciences of the Ancients
by David Hatcher Childress

Childress looks at the technology that was allegedly used in Atlantis and the theory that the Great Pyramid of Egypt was originally a gigantic power station. He examines tales of ancient flight and the technology that it involved; how the ancients used electricity; megalithic building techniques; the use of crystal lenses and the fire from the gods; evidence of various high tech weapons in the past, including atomic weapons; ancient metallurgy and heavy machinery; the role of modern inventors such as Nikola Tesla in bringing ancient technology back into modern use; impossible artifacts; and more.
356 PAGES. 6x9 PAPERBACK. ILLUSTRATED. BIBLIOGRAPHY. $16.95. CODE: TGOD

VIMANA AIRCRAFT OF ANCIENT INDIA & ATLANTIS
by David Hatcher Childress, introduction by Ivan T. Sanderson

In this incredible volume on ancient India, authentic Indian texts such as the *Ramayana* and the *Mahabharata* are used to prove that ancient aircraft were in use more than four thousand years ago. Included in this book is the entire Fourth Century BC manuscript *Vimaanika Shastra* by the ancient author Maharishi Bharadwaaja. Also included are chapters on Atlantean technology, the incredible Rama Empire of India and the devastating wars that destroyed it.
334 PAGES. 6x9 PAPERBACK. ILLUSTRATED. $15.95. CODE: VAA

LOST CONTINENTS & THE HOLLOW EARTH
I Remember Lemuria and the Shaver Mystery
by David Hatcher Childress & Richard Shaver

Shaver's rare 1948 book *I Remember Lemuria* is reprinted in its entirety, and the book is packed with illustrations from Ray Palmer's *Amazing Stories* magazine of the 1940s. Palmer and Shaver told of tunnels running through the earth—tunnels inhabited by the Deros and Teros, humanoids from an ancient spacefaring race that had inhabited the earth, eventually going underground, hundreds of thousands of years ago. Childress discusses the famous hollow earth books and delves deep into whatever reality may be behind the stories of tunnels in the earth. Operation High Jump to Antarctica in 1947 and Admiral Byrd's bizarre statements, tunnel systems in South America and Tibet, the underground world of Agartha, the belief of UFOs coming from the South Pole, more.
344 PAGES. 6x9 PAPERBACK. ILLUSTRATED. $16.95. CODE: LCHE

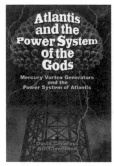

ATLANTIS & THE POWER SYSTEM OF THE GODS
by David Hatcher Childress and Bill Clendenon
Childress' fascinating analysis of Nikola Tesla's broadcast system in light of Edgar Cayce's "Terrible Crystal" and the obelisks of ancient Egypt and Ethiopia. Includes: Atlantis and its crystal power towers that broadcast energy; how these incredible power stations may still exist today; inventor Nikola Tesla's nearly identical system of power transmission; Mercury Proton Gyros and mercury vortex propulsion; more. Richly illustrated, and packed with evidence that Atlantis not only existed—it had a world-wide energy system more sophisticated than ours today.
246 PAGES. 6X9 PAPERBACK. ILLUSTRATED. $15.95. CODE: APSG

THE ANTI-GRAVITY HANDBOOK
edited by David Hatcher Childress
The new expanded compilation of material on Anti-Gravity, Free Energy, Flying Saucer Propulsion, UFOs, Suppressed Technology, NASA Cover-ups and more. Highly illustrated with patents, technical illustrations and photos. This revised and expanded edition has more material, including photos of Area 51, Nevada, the government's secret testing facility. This classic on weird science is back in a new format!
230 PAGES. 7X10 PAPERBACK. ILLUSTRATED. $16.95. CODE: AGH

ANTI-GRAVITY & THE WORLD GRID
Is the earth surrounded by an intricate electromagnetic grid network offering free energy? This compilation of material on ley lines and world power points contains chapters on the geography, mathematics, and light harmonics of the earth grid. Learn the purpose of ley lines and ancient megalithic structures located on the grid. Discover how the grid made the Philadelphia Experiment possible. Explore the Coral Castle and many other mysteries, including acoustic levitation, Tesla Shields and scalar wave weaponry. Browse through the section on anti-gravity patents, and research resources.
274 PAGES. 7X10 PAPERBACK. ILLUSTRATED. $14.95. CODE: AGW

ANTI-GRAVITY & THE UNIFIED FIELD
edited by David Hatcher Childress
Is Einstein's Unified Field Theory the answer to all of our energy problems? Explored in this compilation of material is how gravity, electricity and magnetism manifest from a unified field around us. Why artificial gravity is possible; secrets of UFO propulsion; free energy; Nikola Tesla and anti-gravity airships of the 20s and 30s; flying saucers as superconducting whirls of plasma; anti-mass generators; vortex propulsion; suppressed technology; government cover-ups; gravitational pulse drive; spacecraft & more.
240 PAGES. 7X10 PAPERBACK. ILLUSTRATED. $14.95. CODE: AGU

THE TIME TRAVEL HANDBOOK
A Manual of Practical Teleportation & Time Travel
edited by David Hatcher Childress
The Time Travel Handbook takes the reader beyond the government experiments and deep into the uncharted territory of early time travellers such as Nikola Tesla and Guglielmo Marconi and their alleged time travel experiments, as well as the Wilson Brothers of EMI and their connection to the Philadelphia Experiment—the U.S. Navy's forays into invisibility, time travel, and teleportation. Childress looks into the claims of time travelling individuals, and investigates the unusual claim that the pyramids on Mars were built in the future and sent back in time. A highly visual, large format book, with patents, photos and schematics. Be the first on your block to build your own time travel device!
316 PAGES. 7X10 PAPERBACK. ILLUSTRATED. $16.95. CODE: TTH

MAPS OF THE ANCIENT SEA KINGS
Evidence of Advanced Civilization in the Ice Age
by Charles H. Hapgood

Charles Hapgood has found the evidence in the Piri Reis Map that shows Antarctica, the Hadji Ahmed map, the Oronteus Finaeus and other amazing maps. Hapgood concluded that these maps were made from more ancient maps from the various ancient archives around the world, now lost. Not only were these unknown people more advanced in mapmaking than any people prior to the 18th century, it appears they mapped all the continents. The Americas were mapped thousands of years before Columbus. Antarctica was mapped when its coasts were free of ice!

316 PAGES. 7x10 PAPERBACK. ILLUSTRATED. BIBLIOGRAPHY & INDEX. $19.95. CODE: MASK

PATH OF THE POLE
Cataclysmic Pole Shift Geology
by Charles H. Hapgood

Maps of the Ancient Sea Kings author Hapgood's classic book *Path of the Pole* is back in print! Hapgood researched Antarctica, ancient maps and the geological record to conclude that the Earth's crust has slipped on the inner core many times in the past, changing the position of the pole. *Path of the Pole* discusses the various "pole shifts" in Earth's past, giving evidence for each one, and moves on to possible future pole shifts.

356 PAGES. 6x9 PAPERBACK. ILLUSTRATED. $16.95. CODE: POP

SECRETS OF THE HOLY LANCE
The Spear of Destiny in History & Legend
by Jerry E. Smith

Secrets of the Holy Lance traces the Spear from its possession by Constantine, Rome's first Christian Caesar, to Charlemagne's claim that with it he ruled the Holy Roman Empire by Divine Right, and on through two thousand years of kings and emperors, until it came within Hitler's grasp—and beyond! Did it rest for a while in Antarctic ice? Is it now hidden in Europe, awaiting the next person to claim its awesome power? Neither debunking nor worshiping, *Secrets of the Holy Lance* seeks to pierce the veil of myth and mystery around the Spear. Mere belief that it was infused with magic by virtue of its shedding the Savior's blood has made men kings. But what if it's more? What are "the powers it serves"?

312 PAGES. 6x9 PAPERBACK. ILLUSTRATED. BIBLIOGRAPHY. $16.95. CODE: SOHL

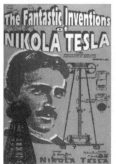

THE FANTASTIC INVENTIONS OF NIKOLA TESLA
by Nikola Tesla with additional material by
David Hatcher Childress

This book is a readable compendium of patents, diagrams, photos and explanations of the many incredible inventions of the originator of the modern era of electrification. In Tesla's own words are such topics as wireless transmission of power, death rays, and radio-controlled airships. In addition, rare material on a secret city built at a remote jungle site in South America by one of Tesla's students, Guglielmo Marconi. Marconi's secret group claims to have built flying saucers in the 1940s and to have gone to Mars in the early 1950s! Incredible photos of these Tesla craft are included. •His plan to transmit free electricity into the atmosphere. •How electrical devices would work using only small antennas. •Why unlimited power could be utilized anywhere on earth. •How radio and radar technology can be used as death-ray weapons in Star Wars.

342 PAGES. 6x9 PAPERBACK. ILLUSTRATED. $16.95. CODE: FINT

REICH OF THE BLACK SUN
Nazi Secret Weapons & the Cold War Allied Legend
by Joseph P. Farrell

Why were the Allies worried about an atom bomb attack by the Germans in 1944? Why did the Soviets threaten to use poison gas against the Germans? Why did Hitler in 1945 insist that holding Prague could win the war for the Third Reich? Why did US General George Patton's Third Army race for the Skoda works at Pilsen in Czechoslovakia instead of Berlin? Why did the US Army not test the uranium atom bomb it dropped on Hiroshima? Why did the Luftwaffe fly a non-stop round trip mission to within twenty miles of New York City in 1944? *Reich of the Black Sun* takes the reader on a scientific-historical journey in order to answer these questions. Arguing that Nazi Germany actually won the race for the atom bomb in late 1944,

352 PAGES. 6x9 PAPERBACK. ILLUSTRATED. BIBLIOGRAPHY. $16.95. CODE: ROBS

THE GIZA DEATH STAR
The Paleophysics of the Great Pyramid & the Military Complex at Giza
by Joseph P. Farrell

Was the Giza complex part of a military installation over 10,000 years ago? Chapters include: An Archaeology of Mass Destruction, Thoth and Theories; The Machine Hypothesis; Pythagoras, Plato, Planck, and the Pyramid; The Weapon Hypothesis; Encoded Harmonics of the Planck Units in the Great Pyramid; High Fregquency Direct Current "Impulse" Technology; The Grand Gallery and its Crystals: Gravito-acoustic Resonators; The Other Two Large Pyramids; the "Causeways," and the "Temples"; A Phase Conjugate Howitzer; Evidence of the Use of Weapons of Mass Destruction in Ancient Times; more.

290 PAGES. 6x9 PAPERBACK. ILLUSTRATED. $16.95. CODE: GDS

THE GIZA DEATH STAR DEPLOYED
The Physics & Engineering of the Great Pyramid
by Joseph P. Farrell

Farrell expands on his thesis that the Great Pyramid was a maser, designed as a weapon and eventually deployed—with disastrous results to the solar system. Includes: Exploding Planets: A Brief History of the Exoteric and Esoteric Investigations of the Great Pyramid; No Machines, Please!; The Stargate Conspiracy; The Scalar Weapons; Message or Machine?; A Tesla Analysis of the Putative Physics and Engineering of the Giza Death Star; Cohering the Zero Point, Vacuum Energy, Flux: Feedback Loops and Tetrahedral Physics; and more.

290 PAGES. 6x9 PAPERBACK. ILLUSTRATED. $16.95. CODE: GDSD

THE GIZA DEATH STAR DESTROYED
The Ancient War For Future Science
by Joseph P. Farrell

Farrell moves on to events of the final days of the Giza Death Star and its awesome power. These final events, eventually leading up to the destruction of this giant machine, are dissected one by one, leading us to the eventual abandonment of the Giza Military Complex—an event that hurled civilization back into the Stone Age. Chapters include: The Mars-Earth Connection; The Lost "Root Races" and the Moral Reasons for the Flood; The Destruction of Krypton: The Electrodynamic Solar System, Exploding Planets and Ancient Wars; Turning the Stream of the Flood: the Origin of Secret Societies and Esoteric Traditions; The Quest to Recover Ancient Mega-Technology; Non-Equilibrium Paleophysics; Monatomic Paleophysics; Frequencies, Vortices and Mass Particles; "Acoustic" Intensity of Fields; The Pyramid of Crystals; tons more.

292 pages. 6x9 paperback. Illustrated. $16.95. Code: GDES

THE TESLA PAPERS
Nikola Tesla on Free Energy &
Wireless Transmission of Power
by Nikola Tesla, edited by David Hatcher Childress

David Hatcher Childress takes us into the incredible world of Nikola Tesla and his amazing inventions. Tesla's fantastic vision of the future, including wireless power, anti-gravity, free energy and highly advanced solar power. Also included are some of the papers, patents and material collected on Tesla at the Colorado Springs Tesla Symposiums, including papers on: •The Secret History of Wireless Transmission •Tesla and the Magnifying Transmitter •Design and Construction of a Half-Wave Tesla Coil •Electrostatics: A Key to Free Energy •Progress in Zero-Point Energy Research •Electromagnetic Energy from Antennas to Atoms •Tesla's Particle Beam Technology •Fundamental Excitatory Modes of the Earth-Ionosphere Cavity

325 PAGES. 8x10 PAPERBACK. ILLUSTRATED. $16.95. CODE: TTP

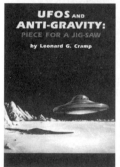

UFOS AND ANTI-GRAVITY
Piece For A Jig-Saw
by Leonard G. Cramp

Leonard G. Cramp's 1966 classic book on flying saucer propulsion and suppressed technology is a highly technical look at the UFO phenomena by a trained scientist. Cramp first introduces the idea of 'anti-gravity' and introduces us to the various theories of gravitation. He then examines the technology necessary to build a flying saucer and examines in great detail the technical aspects of such a craft. Cramp's book is a wealth of material and diagrams on flying saucers, anti-gravity, suppressed technology, G-fields and UFOs. Chapters include Crossroads of Aerodynamics, Aerodynamic Saucers, Limitations of Rocketry, Gravitation and the Ether, Gravitational Spaceships, G-Field Lift Effects, The Bi-Field Theory, VTOL and Hovercraft, Analysis of UFO photos, more.

388 PAGES. 6x9 PAPERBACK. ILLUSTRATED. $16.95. CODE: UAG

THE COSMIC MATRIX
Piece for a Jig-Saw, Part Two
by Leonard G. Cramp

Cramp examines anti-gravity effects and theorizes that this super-science used by the craft—described in detail in the book—can lift mankind into a new level of technology, transportation and understanding of the universe. The book takes a close look at gravity control, time travel, and the interlocking web of energy between all planets in our solar system with Leonard's unique technical diagrams. A fantastic voyage into the present and future!

364 PAGES. 6x9 PAPERBACK. ILLUSTRATED. BIBLIOGRAPHY. $16.00. CODE: CMX

THE A.T. FACTOR
A Scientists Encounter with UFOs
by Leonard Cramp

British aerospace engineer Cramp began much of the scientific anti-gravity and UFO propulsion analysis back in 1955 with his landmark book *Space, Gravity & the Flying Saucer* (out-of-print and rare). In this final book, Cramp brings to a close his detailed and controversial study of UFOs and Anti-Gravity.

324 PAGES. 6x9 PAPERBACK. ILLUSTRATED. BIBLIOGRAPHY. INDEX. $16.95. CODE: ATF

THE FREE-ENERGY DEVICE HANDBOOK
A Compilation of Patents and Reports
by David Hatcher Childress

A large-format compilation of various patents, papers, descriptions and diagrams concerning free-energy devices and systems. *The Free-Energy Device Handbook* is a visual tool for experimenters and researchers into magnetic motors and other "over-unity" devices. With chapters on the Adams Motor, the Hans Coler Generator, cold fusion, superconductors, "N" machines, space-energy generators, Nikola Tesla, T. Townsend Brown, and the latest in free-energy devices. Packed with photos, technical diagrams, patents and fascinating information, this book belongs on every science shelf.

292 PAGES. 8x10 PAPERBACK. ILLUSTRATED. $16.95. CODE: FEH

THE ENERGY GRID
Harmonic 695, The Pulse of the Universe
by Captain Bruce Cathie

This is the breakthrough book that explores the incredible potential of the Energy Grid and the Earth's Unified Field all around us. Cathie's first book, *Harmonic 33*, was published in 1968 when he was a commercial pilot in New Zealand. Since then, Captain Bruce Cathie has been the premier investigator into the amazing potential of the infinite energy that surrounds our planet every microsecond. Cathie investigates the Harmonics of Light and how the Energy Grid is created. In this amazing book are chapters on UFO Propulsion, Nikola Tesla, Unified Equations, the Mysterious Aerials, Pythagoras & the Grid, Nuclear Detonation and the Grid, Maps of the Ancients, an Australian Stonehenge examined, more.

255 PAGES. 6x9 TRADEPAPER. ILLUSTRATED. $15.95. CODE: TEG

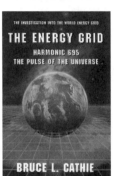

THE BRIDGE TO INFINITY
Harmonic 371244
by Captain Bruce Cathie

Cathie has popularized the concept that the earth is crisscrossed by an electromagnetic grid system that can be used for anti-gravity, free energy, levitation and more. The book includes a new analysis of the harmonic nature of reality, acoustic levitation, pyramid power, harmonic receiver towers and UFO propulsion. It concludes that today's scientists have at their command a fantastic store of knowledge with which to advance the welfare of the human race.

204 PAGES. 6x9 TRADEPAPER. ILLUSTRATED. $14.95. CODE: BTF

THE HARMONIC CONQUEST OF SPACE
by Captain Bruce Cathie

Chapters include: Mathematics of the World Grid; the Harmonics of Hiroshima and Nagasaki; Harmonic Transmission and Receiving; the Link Between Human Brain Waves; the Cavity Resonance between the Earth; the Ionosphere and Gravity; Edgar Cayce—the Harmonics of the Subconscious; Stonehenge; the Harmonics of the Moon; the Pyramids of Mars; Nikola Tesla's Electric Car; the Robert Adams Pulsed Electric Motor Generator; Harmonic Clues to the Unified Field; and more. Also included are tables showing the harmonic relations between the earth's magnetic field, the speed of light, and anti-gravity/gravity acceleration at different points on the earth's surface. New chapters in this edition on the giant stone spheres of Costa Rica, Atomic Tests and Volcanic Activity, and a chapter on Ayers Rock analysed with Stone Mountain, Georgia.

248 PAGES. 6x9. PAPERBACK. ILLUSTRATED. BIBLIOGRAPHY. $16.95. CODE: HCS

GRAVITATIONAL MANIPULATION OF DOMED CRAFT
UFO Propulsion Dynamics
by Paul E. Potter

Potter's precise and lavish illustrations allow the reader to enter directly into the realm of the advanced technological engineer and to understand, quite straightforwardly, the aliens' methods of energy manipulation: their methods of electrical power generation; how they purposely designed their craft to employ the kinds of energy dynamics that are exclusive to space (discoverable in our astrophysics) in order that their craft may generate both attractive and repulsive gravitational forces; their control over the mass-density matrix surrounding their craft enabling them to alter their physical dimensions and even manufacture their own frame of reference in respect to time. Includes a 16-page color insert.

624 pages. 7x10 Paperback. Illustrated. References. $24.00. Code: GMDC

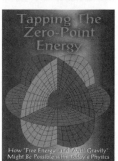

TAPPING THE ZERO POINT ENERGY
Free Energy & Anti-Gravity in Today's Physics
by Moray B. King

King explains how free energy and anti-gravity are possible. The theories of the zero point energy maintain there are tremendous fluctuations of electrical field energy imbedded within the fabric of space. This book tells how, in the 1930s, inventor T. Henry Moray could produce a fifty kilowatt "free energy" machine; how an electrified plasma vortex creates anti-gravity; how the Pons/Fleischmann "cold fusion" experiment could produce tremendous heat without fusion; and how certain experiments might produce a gravitational anomaly.

180 PAGES. 5x8 PAPERBACK. ILLUSTRATED. $12.95. CODE: TAP

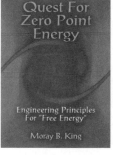

QUEST FOR ZERO-POINT ENERGY
Engineering Principles for "Free Energy"
by Moray B. King

King expands, with diagrams, on how free energy and anti-gravity are possible. The theories of zero point energy maintain there are tremendous fluctuations of electrical field energy embedded within the fabric of space. King explains the following topics: TFundamentals of a Zero-Point Energy Technology; Vacuum Energy Vortices; The Super Tube; Charge Clusters: The Basis of Zero-Point Energy Inventions; Vortex Filaments, Torsion Fields and the Zero-Point Energy; Transforming the Planet with a Zero-Point Energy Experiment; Dual Vortex Forms: The Key to a Large Zero-Point Energy Coherence. Packed with diagrams, patents and photos.

224 PAGES. 6x9 PAPERBACK. ILLUSTRATED. $14.95. CODE: QZPE

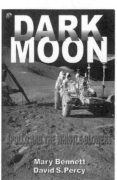

DARK MOON
Apollo and the Whistleblowers
by Mary Bennett and David Percy

Did you know a second craft was going to the Moon at the same time as Apollo 11? Do you know that potentially lethal radiation is prevalent throughout deep space? Do you know there are serious discrepancies in the account of the Apollo 13 'accident'? Did you know that 'live' color TV from the Moon was not actually live at all? Did you know that the Lunar Surface Camera had no viewfinder? Do you know that lighting was used in the Apollo photographs—yet no lighting equipment was taken to the Moon? All these questions, and more, are discussed in great detail by British researchers Bennett and Percy in *Dark Moon*, the definitive book (nearly 600 pages) on the possible faking of the Apollo Moon missions. Tons of NASA photos analyzed for possible deceptions.

568 PAGES. 6x9 PAPERBACK. ILLUSTRATED. BIBLIOGRAPHY. INDEX. $32.00. CODE: DMO

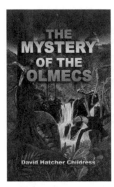

THE MYSTERY OF THE OLMECS
by David Hatcher Childress
The Olmecs were not acknowledged to have existed as a civilization until an international archeological meeting in Mexico City in 1942. Now, the Olmecs are slowly being recognized as the Mother Culture of Mesoamerica, having invented writing, the ball game and the "Mayan" Calendar. But who were the Olmecs? Where did they come from? What happened to them? How sophisticated was their culture? Why are many Olmec statues and figurines seemingly of foreign peoples such as Africans, Europeans and Chinese? Is there a link with Atlantis? In this heavily illustrated book, join Childress in search of the lost cities of the Olmecs! Chapters include: The Mystery of Quizuo; The Mystery of Transoceanic Trade; The Mystery of Cranial Deformation; more.
296 PAGES. 6x9 PAPERBACK. ILLUSTRATED. BIBLIOGRAPHY. COLOR SECTION. $20.00. CODE: MOLM

THE LAND OF OSIRIS
An Introduction to Khemitology
by Stephen S. Mehler
Was there an advanced prehistoric civilization in ancient Egypt who built the great pyramids and carved the Great Sphinx? Did the pyramids serve as energy devices and not as tombs for kings? Mehler has uncovered an indigenous oral tradition that still exists in Egypt, and has been fortunate to have studied with a living master of this tradition, Abd'El Hakim Awyan. Mehler has also been given permission to present these teachings to the Western world, teachings that unfold a whole new understanding of ancient Egypt . Chapters include: Egyptology and Its Paradigms; Asgat Nefer—The Harmony of Water; Khemit and the Myth of Atlantis; The Extraterrestrial Question; more.
272 PAGES. 6x9 PAPERBACK. ILLUSTRATED. COLOR SECTION. BIBLIOGRAPHY. $18.00 CODE: LOOS

ABOMINABLE SNOWMEN:
LEGEND COME TO LIFE
The Story of Sub-Humans on Six Continents from the Early Ice Age Until Today
by Ivan T. Sanderson
Do "Abominable Snowmen" exist? Prepare yourself for a shock. In the opinion of one of the world's leading naturalists, not one, but possibly four kinds, still walk the earth! Do they really live on the fringes of the towering Himalayas and the edge of myth-haunted Tibet? From how many areas in the world have factual reports of wild, strange, hairy men emanated? Reports of strange apemen have come in from every continent, except Antarctica.
525 PAGES. 6x9 PAPERBACK. ILLUSTRATED. BIBLIOGRAPHY. INDEX. $16.95. CODE: ABML

INVISIBLE RESIDENTS
The Reality of Underwater UFOS
by Ivan T. Sanderson
In this book, Sanderson, a renowned zoologist with a keen interest in the paranormal, puts forward the curious theory that "OINTS"—Other Intelligences—live under the Earth's oceans. This underwater, parallel, civilization may be twice as old as Homo sapiens, he proposes, and may have "developed what we call space flight." Sanderson postulates that the OINTS are behind many UFO sightings as well as the mysterious disappearances of aircraft and ships in the Bermuda Triangle. What better place to have an impenetrable base than deep within the oceans of the planet? Sanderson offers here an exhaustive study of USOs (Unidentified Submarine Objects) observed in nearly every part of the world.
298 PAGES. 6x9 PAPERBACK. ILLUSTRATED. BIBLIOGRAPHY. INDEX. $16.95. CODE: INVS

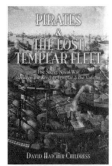

PIRATES & THE LOST TEMPLAR FLEET
The Secret Naval War Between the Templars & the Vatican
by David Hatcher Childress

Childress takes us into the fascinating world of maverick sea captains who were Knights Templar (and later Scottish Rite Free Masons) who battled the ships that sailed for the Pope. The lost Templar fleet was originally based at La Rochelle in southern France, but fled to the deep fiords of Scotland upon the dissolution of the Order by King Phillip. This banned fleet of ships was later commanded by the the St. Clair family of Rosslyn Chapel (birthplace of Free Masonry). St. Clair and his Templars made a voyage to Canada in the year 1298 AD, nearly 100 years before Columbus! Later, this fleet of ships and new ones to come, flew the Skull and Crossbones, the symbol of the Knights Templar.

320 PAGES. 6x9 PAPERBACK. ILLUSTRATED. BIBLIOGRAPHY. $16.95. CODE: PLTF

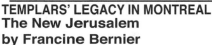

TEMPLARS' LEGACY IN MONTREAL
The New Jerusalem
by Francine Bernier

The book reveals the links between Montreal and: John the Baptist as patron saint; Melchizedek, the first king-priest and a father figure to the Templars and the Essenes; Stella Maris, the Star of the Sea from Mount Carmel; the Phrygian goddess Cybele as the androgynous Mother of the Church; St. Blaise, the Armenian healer or "Therapeut"- the patron saint of the stonemasons and a major figure to the Benedictine Order and the Templars; the presence of two Black Virgins; an intriguing family coat of arms with twelve blue apples; and more.

352 PAGES. 6x9 PAPERBACK. ILLUSTRATED. BIBLIOGRAPHY. $21.95. CODE: TLIM

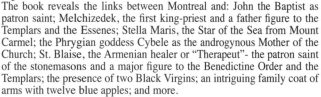

THE HISTORY OF THE KNIGHTS TEMPLARS
by Charles G. Addison, introduction by David Hatcher Childress

Chapters on the origin of the Templars, their popularity in Europe and their rivalry with the Knights of St. John, later to be known as the Knights of Malta. Detailed information on the activities of the Templars in the Holy Land, and the 1312 AD suppression of the Templars in France and other countries, which culminated in the execution of Jacques de Molay and the continuation of the Knights Templars in England and Scotland; the formation of the society of Knights Templars in London; and the rebuilding of the Temple in 1816. Plus a lengthy intro about the lost Templar fleet and its North American sea routes.

395 PAGES. 6x9 PAPERBACK. ILLUSTRATED. $16.95. CODE: HKT

OTTO RAHN AND THE QUEST FOR THE HOLY GRAIL
The Amazing Life of the Real "Indiana Jones"
by Nigel Graddon

Otto Rahn led a life of incredible adventure in southern France in the early 1930s. The Hessian language scholar is said to have found runic Grail tablets in the Pyrenean grottoes, and decoded hidden messages within the medieval Grail masterwork *Parsifal*. The fabulous artifacts identified by Rahn were believed by Himmler to include the Grail Cup, the Spear of Destiny, the Tablets of Moses, the Ark of the Covenant, the Sword and Harp of David, the Sacred Candelabra and the Golden Urn of Manna. Some believe that Rahn was a Nazi guru who wielded immense influence on his elders and "betters" within the Hitler regime, persuading them that the Grail was the Sacred Book of the Aryans, which, once obtained, would justify their extreme political theories and revivify the ancient Germanic myths. But things are never as they seem, and as new facts emerge about Otto Rahn a far more extraordinary story unfolds.

450 pages. 6x9 Paperback. Illustrated. Appendix. Index. $18.95. Code: ORQG

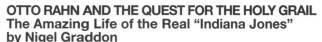

ANCIENT ALIENS ON THE MOON
By Mike Bara
What did NASA find in their explorations of the solar system that they may have kept from the general public? How ancient really are these ruins on the Moon? Using official NASA and Russian photos of the Moon, Bara looks at vast cityscapes and domes in the Sinus Medii region as well as glass domes in the Crisium region. Bara also takes a detailed look at the mission of Apollo 17 and the case that this was a salvage mission, primarily concerned with investigating an opening into a massive hexagonal ruin near the landing site. Chapters include: The History of Lunar Anomalies; The Early 20th Century; Sinus Medii; To the Moon Alice!; Mare Crisium; Yes, Virginia, We Really Went to the Moon; Apollo 17; more. Tons of photos of the Moon examined for possible structures and other anomalies.
248 Pages. 6x9 Paperback. Illustrated.. $19.95. Code: AAOM

ANCIENT ALIENS ON MARS
By Mike Bara
Bara brings us this lavishly illustrated volume on alien structures on Mars. Was there once a vast, technologically advanced civilization on Mars, and did it leave evidence of its existence behind for humans to find eons later? Did these advanced extraterrestrial visitors vanish in a solar system wide cataclysm of their own making, only to make their way to Earth and start anew? Was Mars once as lush and green as the Earth, and teeming with life? Chapters include: War of the Worlds; The Mars Tidal Model; The Death of Mars; Cydonia and the Face on Mars; The Monuments of Mars; The Search for Life on Mars; The True Colors of Mars and The Pathfinder Sphinx; more. Color section.
252 Pages. 6x9 Paperback. Illustrated. $19.95. Code: AMAR

ANCIENT ALIENS ON MARS II
By Mike Bara
Using data acquired from sophisticated new scientific instruments like the Mars Odyssey THEMIS infrared imager, Bara shows that the region of Cydonia overlays a vast underground city full of enormous structures and devices that may still be operating. He peels back the layers of mystery to show images of tunnel systems, temples and ruins, and exposes the sophisticated NASA conspiracy designed to hide them. Bara also tackles the enigma of Mars' hollowed out moon Phobos, and exposes evidence that it is artificial. Long-held myths about Mars, including claims that it is protected by a sophisticated UFO defense system, are examined. Data from the Mars rovers Spirit, Opportunity and Curiosity are examined; everything from fossilized plants to mechanical debris is exposed in images taken directly from NASA's own archives.
294 Pages. 6x9 Paperback. Illustrated. $19.95. Code: AAM2

ANCIENT TECHNOLOGY IN PERU & BOLIVIA
By David Hatcher Childress
Childress speculates on the existence of a sunken city in Lake Titicaca and reveals new evidence that the Sumerians may have arrived in South America 4,000 years ago. He demonstrates that the use of "keystone cuts" with metal clamps poured into them to secure megalithic construction was an advanced technology used all over the world, from the Andes to Egypt, Greece and Southeast Asia. He maintains that only power tools could have made the intricate articulation and drill holes found in extremely hard granite and basalt blocks in Bolivia and Peru, and that the megalith builders had to have had advanced methods for moving and stacking gigantic blocks of stone, some weighing over 100 tons.
340 Pages. 6x9 Paperback. Illustrated.. $19.95 Code: ATP

THE CRYSTAL SKULLS
Astonishing Portals to Man's Past
by David Hatcher Childress and Stephen S. Mehler

Childress introduces the technology and lore of crystals, and then plunges into the turbulent times of the Mexican Revolution form the backdrop for the rollicking adventures of Ambrose Bierce, the renowned journalist who went missing in the jungles in 1913, and F.A. Mitchell-Hedges, the notorious adventurer who emerged from the jungles with the most famous of the crystal skulls. Mehler shares his extensive knowledge of and experience with crystal skulls. Having been involved in the field since the 1980s, he has personally examined many of the most influential skulls, and has worked with the leaders in crystal skull research, including the inimitable Nick Nocerino, who developed a meticulous methodology for the purpose of examining the skulls.
294 pages. 6x9 Paperback. Illustrated. Bibliography. $18.95. Code: CRSK

THE INCREDIBLE LIGHT BEINGS OF THE COSMOS
Are Orbs Intelligent Light Beings from the Cosmos?
by Antonia Scott-Clark

Scott-Clark has experienced orbs for many years, but started photographing them in earnest in the year 2000 when the "Light Beings" entered her life. She took these very seriously and set about privately researching orb occurrences. The incredible results of her findings are presented here, along with many of her spectacular photographs. With her friend, GoGos lead singer Belinda Carlisle, Antonia tells of her many adventures with orbs. Find the answers to questions such as: Can you see orbs with the naked eye?; Are orbs intelligent?; What are the Black Villages?; What is the connection between orbs and crop circles? Antonia gives detailed instruction on how to photograph orbs, and how to communicate with these Light Beings of the Cosmos.
334 pages. 6x9 Paperback. Illustrated. References. $19.95. Code: ILBC

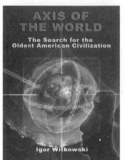

AXIS OF THE WORLD
The Search for the Oldest American Civilization
by Igor Witkowski

Polish author Witkowski's research reveals remnants of a high civilization that was able to exert its influence on almost the entire planet, and did so with full consciousness. Sites around South America show that this was not just one of the places influenced by this culture, but a place where they built their crowning achievements. Easter Island, in the southeastern Pacific, constitutes one of them. The Rongo-Rongo language that developed there points westward to the Indus Valley. Taken together, the facts presented by Witkowski provide a fresh, new proof that an antediluvian, great civilization flourished several millennia ago.
220 pages. 6x9 Paperback. Illustrated. References. $18.95. Code: AXOW

LEY LINE & EARTH ENERGIES
An Extraordinary Journey into the Earth's Natural Energy System
by David Cowan & Chris Arnold

The mysterious standing stones, burial grounds and stone circles that lace Europe, the British Isles and other areas have intrigued scientists, writers, artists and travellers through the centuries. How do ley lines work? How did our ancestors use Earth energy to map their sacred sites and burial grounds? How do ghosts and poltergeists interact with Earth energy? How can Earth spirals and black spots affect our health? This exploration shows how natural forces affect our behavior, how they can be used to enhance our health and well being.
368 PAGES. 6x9 PAPERBACK. ILLUSTRATED. $18.95. CODE: LLEE

ROSWELL AND THE REICH
The Nazi Connection
By Joseph P. Farrell
Farrell has meticulously reviewed the best-known Roswell research from UFO-ET advocates and skeptics alike, as well as some little-known source material, and comes to a radically different scenario of what happened in Roswell, New Mexico in July 1947, and why the US military has continued to cover it up to this day. Farrell presents a fascinating case sure to disturb both ET believers and disbelievers, namely, that what crashed may have been representative of an independent postwar Nazi power—an extraterritorial Reich monitoring its old enemy, America, and the continuing development of the very technologies confiscated from Germany at the end of the War.
540 pages. 6x9 Paperback. Illustrated. $19.95. Code: RWR

SECRETS OF THE UNIFIED FIELD
The Philadelphia Experiment, the Nazi Bell, and the Discarded Theory
by Joseph P. Farrell
Farrell examines the now discarded Unified Field Theory. American and German wartime scientists and engineers determined that, while the theory was incomplete, it could nevertheless be engineered. Chapters include: The Meanings of "Torsion"; Wringing an Aluminum Can; The Mistake in Unified Field Theories and Their Discarding by Contemporary Physics; Three Routes to the Doomsday Weapon: Quantum Potential, Torsion, and Vortices; Tesla's Meeting with FDR; Arnold Sommerfeld and Electromagnetic Radar Stealth; Electromagnetic Phase Conjugations, Phase Conjugate Mirrors, and Templates; The Unified Field Theory, the Torsion Tensor, and Igor Witkowski's Idea of the Plasma Focus; tons more.
340 pages. 6x9 Paperback. Illustrated. $18.95. Code: SOUF

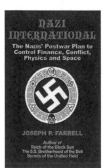

NAZI INTERNATIONAL
The Nazi's Postwar Plan to Control Finance, Conflict, Physics and Space
by Joseph P. Farrell
Beginning with prewar corporate partnerships in the USA, including some with the Bush family, he moves on to the surrender of Nazi Germany, and evacuation plans of the Germans. He then covers the vast, and still-little-known recreation of Nazi Germany in South America with help of Juan Peron, I.G. Farben and Martin Bormann. Farrell then covers Nazi Germany's penetration of the Muslim world including Wilhelm Voss and Otto Skorzeny in Gamel Abdul Nasser's Egypt before moving on to the development and control of new energy technologies including the Bariloche Fusion Project, Dr. Philo Farnsworth's Plasmator, and the work of Dr. Nikolai Kozyrev. Finally, Farrell discusses the Nazi desire to control space, and examines their connection with NASA, the esoteric meaning of NASA Mission Patches.
412 pages. 6x9 Paperback. Illustrated. $19.95. Code: NZIN

ARKTOS
The Polar Myth in Science, Symbolism & Nazi Survival
by Joscelyn Godwin
Explored are the many tales of an ancient race said to have lived in the Arctic regions, such as Thule and Hyperborea. Progressing onward, he looks at modern polar legends: including the survival of Hitler, German bases in Antarctica, UFOs, the hollow earth, and the hidden kingdoms of Agartha and Shambala. Chapters include: Prologue in Hyperborea; The Golden Age; The Northern Lights; The Arctic Homeland; The Aryan Myth; The Thule Society; The Black Order; The Hidden Lands; Agartha and the Polaires; Shambhala; The Hole at the Pole; Antarctica; more.
220 Pages. 6x9 Paperback. Illustrated. Bib. Index. $16.95. Code: ARK

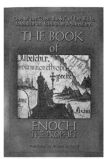

THE BOOK OF ENOCH
translated by Richard Laurence
This is a reprint of the Apocryphal *Book of Enoch the Prophet* which was first discovered in Abyssinia in the year 1773 by a Scottish explorer named James Bruce. One of the main influences from the book is its explanation of evil coming into the world with the arrival of the "fallen angels." Enoch acts as a scribe, writing up a petition on behalf of these fallen angels, or fallen ones, to be given to a higher power for ultimate judgment. Christianity adopted some ideas from Enoch, including the Final Judgment, the concept of demons, the origins of evil and the fallen angels, and the coming of a Messiah and ultimately, a Messianic kingdom.
224 PAGES. 6x9 PAPERBACK. ILLUSTRATED. INDEX. $16.95. CODE: BOE

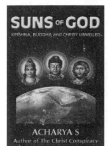

SUNS OF GOD
Krishna, Buddha and Christ Unveiled
by Acharya S
Over the past several centuries, the Big Three spiritual leaders have been the Lords Christ, Krishna and Buddha, whose stories and teachings are so remarkably similar as to confound and amaze those who encounter them. As classically educated archaeologist, historian, mythologist and linguist Acharya S thoroughly reveals, these striking parallels exist not because these godmen were "historical" personages who "walked the earth" but because they are personifications of the central focus of the famous and scandalous "mysteries." These mysteries date back thousands of years and are found globally, reflecting an ancient tradition steeped in awe and intrigue.
428 PAGES. 6x9 PAPERBACK. ILLUSTRATED. BIBLIOGRAPHY. INDEX. $18.95. CODE: SUNG

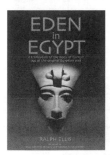

EDEN IN EGYPT
by Ralph Ellis
The story of Adam and Eve from the Book of Genesis is perhaps one of the best-known stories in circulation, even today, and yet nobody really knows where this tale came from or what it means. But even a cursory glance at the text will demonstrate the origins of this tale, for the river of Eden is described as having four branches. There is only one river in this part of the world that fits this description, and that is the Nile, with the four branches forming the Nile Delta. According to Ellis, Judaism was based upon the reign of the pharaoh Akhenaton, because the solitary Judaic god was known as Adhon while this pharaoh's solitary god was called Aton or Adjon. But what of the identities of Adam and Eve? Includes 16 page color section.
320 PAGES. 6x9 PAPERBACK. ILLUSTRATED. BIBLIOGRAPHY. INDEX. $20.00. CODE: EIE

ELVIS IS ALIVE
The Complete Conspiracy
By Xaviant Haze
Haze blows the Elvis conspiracies vault wide open with the first book dedicated to the mysteries surrounding the King of Rock n' Roll. His was a tale of screaming fans, untold riches and a spectacular free fall into obliteration. It was also a tale of UFO encounters, zany fake funerals, studies in metaphysics, numerology, occult theology—and a strange connection with Michael Jackson. An infamous pill addiction led the King to an early "death" and in the aftermath a pop culture phenomenon was born, enabling Elvis to sustain a famous afterlife thanks to over three decades of conspiracy theories. Includes the complete comic strip "Elvis Presley: His Story in Pictures." Contents include: Pills and Cheeseburgers; Elvis Dies!; Elvis Lives!; Elvis versus the Mafia; Elvis the Esoteric; The Michael Jackson Connection; Long Live the King; more. Eight-page color section.
204 Pages. 6x9 Paperback. Illustrated. $19.95 Code: ELVA

ORDER FORM

**10% Discount
When You Order
3 or More Items!**

One Adventure Place
P.O. Box 74
Kempton, Illinois 60946
United States of America
Tel.: 815-253-6390 • Fax: 815-253-6300
Email: auphq@frontiernet.net
http://www.adventuresunlimitedpress.com

ORDERING INSTRUCTIONS

✓ Remit by USD$ Check, Money Order or Credit Card

✓ Visa, Master Card, Discover & AmEx Accepted

✓ Paypal Payments Can Be Made To:

    info@wexclub.com

✓ Prices May Change Without Notice

✓ 10% Discount for 3 or More Items

SHIPPING CHARGES

United States

✓ Postal Book Rate { $4.50 First Item
  50¢ Each Additional Item

✓ POSTAL BOOK RATE Cannot Be Tracked!
  Not responsible for non-delivery.

✓ Priority Mail { $6.00 First Item
  $2.00 Each Additional Item

✓ UPS { $7.00 First Item
  $1.50 Each Additional Item

  NOTE: UPS Delivery Available to Mainland USA Only

Canada

✓ Postal Air Mail { $15.00 First Item
  $3.00 Each Additional Item

✓ Personal Checks or Bank Drafts MUST BE

  US$ and Drawn on a US Bank

✓ Canadian Postal Money Orders OK

✓ Payment MUST BE US$

All Other Countries

✓ Sorry, No Surface Delivery!

✓ Postal Air Mail { $19.00 First Item
  $7.00 Each Additional Item

✓ Checks and Money Orders MUST BE US$
  and Drawn on a US Bank or branch.

✓ Paypal Payments Can Be Made in US$ To:
  info@wexclub.com

SPECIAL NOTES

✓ RETAILERS: Standard Discounts Available

✓ BACKORDERS: We Backorder all Out-of-
  Stock Items Unless Otherwise Requested

✓ PRO FORMA INVOICES: Available on Request

✓ DVD Return Policy: Replace defective DVDs only

ORDER ONLINE AT: www.adventuresunlimitedpress.com

**10% Discount When You Order
3 or More Items!**

Please check: ✓

☐ This is my first order   ☐ I have ordered before

Name	
Address	
City	

| State/Province | | Postal Code | |

Country	
Phone: Day	Evening
Fax	Email

Item Code	Item Description	Qty	Total

Please check: ✓

	Subtotal ▶	
	Less Discount-10% for 3 or more items ▶	
☐ Postal-Surface	Balance ▶	
☐ Postal-Air Mail Illinois Residents 6.25% Sales Tax ▶		
(Priority in USA)	Previous Credit ▶	
☐ UPS	Shipping ▶	
(Mainland USA only) Total (check/MO in USD$ only) ▶		

☐ Visa/MasterCard/Discover/American Express

Card Number:

Expiration Date: Security Code:

✓ SEND A CATALOG TO A FRIEND: